NATIONAL AGRICULTURAL BIOTECHNOLOGY COUNCIL REPORTS

1992

NABC Report 4

Animal Biotechnology:
Opportunities & Challenges

The National Agricultural Biotechnology Council provides an open forum for the discussion of issues related to the impact of biotechnology on agriculture. The views presented and positions taken by individual participants in this report are their own and do not necessarily reflect the views or policies of the NABC.

Photos on cover and pages 23, 73 & 175 courtesy of University Photography, Cornell. Photo on page 1 courtesy of Agricultural Communications, Texas A&M University.

Additional copies are available for $5
please make checks or purchase orders payable to:

NABC / BTI
159 Biotechnology Building
Cornell University
Ithaca, NY 14853–2703

ISBN: 0–9630907–2–0
Library of Congress Catalog Card Number 92-062575

Printed by Braun & Brumfield, Ann Arbor, Michigan
printed on recycled paper

NABC REPORT 4

Animal Biotechnology: *Opportunities & Challenges*

Edited by June Fessenden MacDonald

published by
NATIONAL AGRICULTURAL BIOTECHNOLOGY COUNCIL
ITHACA, NEW YORK 14853-2703

NATIONAL AGRICULTURAL BIOTECHNOLOGY COUNCIL

Providing an open forum
for exploring issues in
agricultural biotechnology.

NABC is a consortium of not-for-profit agricultural research and educational institutions established in 1988.

member institutions

BOYCE THOMPSON INSTITUTE

CORNELL UNIVERSITY

IOWA STATE UNIVERSITY

MICHIGAN STATE UNIVERSITY

OHIO STATE UNIVERSITY

PURDUE UNIVERSITY

RUTGERS UNIVERSITY

THE TEXAS A&M UNIVERSITY SYSTEM

TUFTS UNIVERSITY

UNIVERSITY OF CALIFORNIA, DAVIS

UNIVERSITY OF GEORGIA

UNIVERSITY OF MISSOURI, COLUMBIA

UNIVERSITY OF NEBRASKA, LINCOLN

Other Reports by NABC:

NABC REPORT 1, Biotechnology and Sustainable Agriculture: Policy Alternatives (1989)

NABC REPORT 2, Agricultural Biotechnology, Food Safety and Nutritional Quality for the Consumer (1990)

NABC REPORT 3, Agricultural Biotechnology at the Crossroads: Biological, Social and Institutional Concerns (1991)

Many contributed to the success of the fourth annual NABC meeting and the production of this report. Special thanks goes to the Planning Committee at the host institution, Texas A&M University: co-chairs Paul Thompson and John Shadduck, with Gary Adams, Floyd Byers, Russell Cross, Nancy Turner, Robert Wells and James Womack, and to Gary Varner who organized the optional seminar on ethics and patenting. NABC gratefully acknowledges the help and support of Charles Arntzen, J. Charles Lee, and the Institute of Biosciences and Technology. Also recognized is the important role of the workshop facilitators: Rick Bennett from the North Bay Public Policy and Education Team, Davis, California; Laura Meagher, Agricultural Biotechnology Center, Cook College, Rutgers University; Ellen Ritter, Agricultural Communications, Texas A&M University; and Barbara Gastel MD, Journalism, Texas A&M University, as well as the work of the several graduate and undergraduate students who assisted in the meetings success.

Special recognition goes to the efficient conference organizational service provided by Barbara Stow, with assistance from Kelly Hancock, of the Center for Biotechnology Policy and Ethics, Texas A&M University. Carolyn Grine of their College of Veterinary Medicine was also most helpful .

Very special thanks and recognition goes to Kate O'Hara, Design/Editorial Coordinator, NABC, who is responsible for the overall design and production of this report, not only for the additional work she had with the editor on sabbatical leave, but also for her general oversight of meeting and report details.

Finally, the cooperation received during the production of this report from the meeting organizers, presenters, workshop chairs and participants was great and is sincerely appreciated.

June Fessenden MacDonald
Deputy Director, NABC
Editor

June Fessenden MacDonald
Deputy Director, NABC
Biochemistry, Molecular and Cell Biology
Cornell University

The National Agricultural Biotechnology Council (NABC), organized in 1988, added 5 new members in 1992, bringing the total membership to 13 not–for–profit agricultural research and educational institutions. NABC's principal objectives are to:

—*Provide an open forum for persons with different interests and concerns to come together to speak, to listen, to learn and to participate in meaningful dialogue and evaluation of the potential impacts of agricultural biotechnology*

—*Define issues and public policy options related to biotechnology in the food, agricultural and environmental areas*

—*Promote increased understanding of the scientific, economic, legislative and social issues associated with agricultural biotechnology by compiling and disseminating information to interested people*

—*Facilitate active communication among researchers, administrators, policymakers, practitioners and other concerned people to insure that all viewpoints contribute to the safe and efficacious development of biotechnology for the benefit of society*

—*Sponsor meetings and workshops and publish and distribute reports that provide a foundation for addressing issues*

The Fourth Annual NABC Meeting (NABC 4), hosted by the Texas A&M University System, once again demonstrated the importance of an open forum where people with different values and strongly held viewpoints can come together as equal participants in a dialogue on agricultural biotechnology. For some participants it was the first time they had personally met someone with strongly held opposing views and truly listened to their arguments.

There was a more diverse group of participants, many attending their first NABC meeting, discussing issues of animal well–being, meat and animal product safety and regulatory policy. The new topic addressed by NABC was the linkages between animal science, veterinary medicine and human medicine and how they could be encouraged since these groups now have limited interaction. While discussions in this area were generally harmonious, the other areas elicited more lively, even contentious, discussions. Still, participants in each workshop were able to reach some areas of consensus. This report, hopefully, communicates some of the flavor of the meeting.

The need for real communication and open procedures was never more evident than at NABC 4. Not only was the call for open dialogue sounded in every session, but also the need for improved communication, especially during policy development. Improved communication rivaled animal well–being as the prime topic during informal discussions. The highly publicized announcement of FDA policy one day before the meeting caused many participants, representing all sides of the animal biotechnology dialogue, to be concerned. Most participants were not against the policy as announced, but rather they were concerned that, regardless of the value of the policy, a process that leaves any stakeholders feeling shut out of the process—and there were several at the meeting—does not enhance the government's or biotechnology's credibility in the eyes of the public.

NABC, in its 5 years of existence, provides opportunity for all stakeholders to come together in a consensus building setting on biotechnology and agriculture. We believe that policies developed with an open forum with input from all interested stakeholders will benefit biotechnology through greater public acceptance of these policies. NABC has accepted the encouragement of communication and open dialogue among all stakeholders in agricultural biotechnology as a central mission.

NABC hopes this report will contribute to increased understanding of the range of viewpoints on animal biotechnology and stimulate improved dialogue, provide new information from different perspectives for all those interested in and affected by animal biotechnology and also provide a foundation for addressing some of the concerns facing society today in this burgeoning area of agricultural biotechnology.

In this single volume, NABC has tried to provide different readers with insight into the opportunities and challenges in animal biotechnology. Part I of this report provides a synopsis of NABC 4, highlighting the issues raised by speakers and participants as well as statements and recommendations on which there was consensus by workshop participants. For those readers who desire more detail, the keynote addresses are found in Part II. Background papers and full workshop reports are combined topically in Part III.

Pleasant reading and productive dialogue.

TABLE OF CONTENTS

PART I ANIMAL BIOTECHNOLOGY: OPPORTUNITIES & CHALLENGES

2

Paul B. Thompson
Director, Center for Biotechnology Policy and Ethics
Texas A&M University

John A. Shadduck
Associate Deputy Chancellor and
Dean, College of Veterinary Medicine
Texas A&M University

Overview

NABC 4, the fourth annual open forum on agricultural biotechnology, was devoted to issues in animal biotechnology. The May 1992 meeting was held in College Station, Texas and hosted by Texas A&M University. Animal well–being, the safety of animal food products and regulatory issues were on the agenda along with the examination of links between animal biotechnology and new opportunities in human and animal medicine. Previous NABC meetings had focused upon sustainability, food safety and quality, and the financial and regulatory prospects for agricultural biotechnology at a crossroads. Animal biotechnology, especially recombinant DNA research on agricultural animals, however, has opened doors to entirely new areas of human endeavor, many of which were not well understood in the past. The NABC 4 presentations and discussions on animal biotechnology revisited several themes that had been discussed at previous NABC meetings, but also broke ground in identifying several key topics that had not been examined at Ames, Iowa in 1989; at Ithaca, New York in 1990; and at Sacramento, California in 1991.

As in previous meetings, the aim of the National Agricultural Biotechnology Council was first and foremost to establish an open setting in which all perspectives and interests can be represented with participants sharing ideas, asking questions and interacting with one another. Convening under the banner *Animal Biotechnology: Opportunities and Challenges*, more than 150 representatives of industry, interest groups, government and universities opened the conversation on these topics with the goals of establishing a common base of knowledge among all participants, reaching consensus where possible, and specifying a limited number of recommendations to emerge as the product of the workshops. NABC 4 continued the trend of the three previous meetings by expanding the range of views and groups represented. Many of the participants, particularly those concerned with issues of animal well–being, were attending this open forum for the first time. Also, some participants were presented for the first time with viewpoints in sharp contrast to their own. For this reason alone, NABC 4 clearly can be said to have served the NABC mission of promoting dialogue among those with different views.

NEW THEMES

The new topic to the NABC forum was the linkages between human medicine and animal agriculture. Although animal biotechnology was addressed at NABC 1 and at NABC 3, animal well–being as a special theme was a new focus for NABC.

Animal Well–Being

The dialogue on animal well–being had been conceptualized as an opportunity to take up the question of whether developments in animal biotechnology would produce any new or unanticipated issues for the well–being of agricultural animals. Although these topics were, indeed, discussed, presentations by keynoter Michael Fox, Vice President for Bioethics and Farm Animals, The Humane Society of the United States; David Meeker, Director of Research and Education, National Pork Producers Council; and Bernard Rollin, Professor of Philosophy at Colorado State University, moved the discussions into more philosophical and broad–ranging areas. This workshop became a forum in which those who saw themselves as representing animals and those who saw themselves as representing agriculture engaged in energetic dialogue over the criteria and basis for extending concern to animals, without respect to whether biotechnology or, indeed, even agriculture was the topic of concern. As such, participants raised examples from human biomedical research and product testing, familiar forums for animal welfare debates, as a means for sounding out each other's basic views on animal issues.

The question of biomedical applications came up particularly with the "new creation" of: 1. precise animal models for human diseases; and 2. animals as "bioreactors" producing human pharmaceuticals. Rollin prompted a discussion of the dilemma of balancing the relief of great human pain and suffering from genetic diseases with the large numbers of animals that would experience great suffering. He suggested researchers could eliminate the pain centers of such animals, but noted that this, too, raises ethical and aesthetic concerns.

Although the heated discussions in this group produced limited consensus, there was general agreement that it is acceptable under conditions where animals do not experience great suffering, to use animals for human use—whether for food production, as "bioreactors," or as research models for improving human and animal health.

The intensity of discussions in the group was evident to all meeting participants and issues of animal well–being wound up being raised (sometimes briefly) in every workshop. In the shadow of such lively dialogue, the workshop group examining links to human health felt itself to be too homogeneous with few issues on which participants opinions diverged. The group invited a participant from the animal well–being workshop to a session to learn, at least, what the hubbub was about. Although the agenda was broader, and the consensus achievements occurred in other areas, the 1992 meeting will undoubtedly be remembered as the "animal welfare" meeting of the National Agricultural Biotechnology Council.

Links to Human Health

Recombinant DNA research on farm animals conducted in animal science departments of agricultural universities and in colleges of veterinary medicine, has begun to bring the scientists in these areas into the prospects and controversies that have traditionally been associated with biomedical research. As highlighted by keynote speaker Neal First, Professor of Animal Science at the University of Wisconsin, animal biotechnology continues to establish breakthroughs in reproductive technology, enhance genetic changes in animals and improve animal health. He conveyed to the participants some of the excitement felt by researchers, himself included, as they push the frontiers of animal science forward. It was noted that basic research aimed at disease control in animals often spills over to human applications. Fuller Bazer, Animal Science Department, Texas A&M University, asserted that "When human and animals have diseases with common etiology and genetic markers of the disease, genetic or therapeutic solutions will favorably impact both human and animal health." What is more, Clifton Baile, Director of Research and Development at Monsanto, suggested that intensive public and private funding for research in biomedical applications such as gene mapping and pharmaceuticals will produce techniques, methods and models that will shorten the time for product development for those working on farm animals.

The workshop participants actively discussed the need for connections, or "new linkages," between human medicine, animal medicine and animal agriculture. They saw a major potential for expanding the dialogue and research interaction among these groups, suggesting that soon the justification for animal biotechnology may be its great benefits to human health, not just animal productivity.

It was, however, noted that as agricultural researchers expand their research and interact directly, or even indirectly, with biomedical researchers, they can expect to face some of the problems that have existed in public health and the biomedical research policy arena for some time. These include an intense public interest in reproductive technologies, in part because of their relevance to the abortion issue, and also a level of public concern for the well–being of animals exceeding that hitherto experienced in connection with food animals. As such, the workshop on animal well–being experienced an overlap with the workshop on animal biotechnology links to human health that meeting organizers had not anticipated. When continuing discussions on food safety and regulatory policy were added into the mix, the two–plus days of the meeting proved stimulating.

REVISITED THEMES

Communication and Open Dialogue

The only issue that could rival animal well–being as a main current at the 1992 meeting was the continued call from participants for open communication and the need for all stakeholders to be involved early and continuously in the dialogue on biotechnology. The highly publicized White House announcement of the FDA's policy for evaluating the safety of foods of plant origin made on May 26, 1992, the day before the meeting began, resulted in several diverse groups in attendance at NABC 4 openly expressing concern that if biotechnology is to gain public acceptance, policies must be developed within an open framework with opportunity for input by all interested stakeholders. While most in attendance were not against the announced policy per se, participants on "both" sides of other biotechnology issues followed press coverage of the announcement all week, expressing the view that the announcement reflected little understanding of how public concerns and questions about biotechnology can be addressed in a manner that inspires confidence in the regulatory process.

At the final plenary session, in response to a recommendation by the participants in the Regulatory Issues workshop, several participants spoke forcefully in favor (no one spoke in opposition) of NABC corresponding with appropriate federal officials urging more open dialogue during future deliberations about agricultural biotechnology policies.[1]

Ironically, the announcement of this policy spoke directly to two of the concerns expressed as major themes of the 1991 NABC meeting in California. There, U.S. competitiveness had been linked to a need for clear delineation of regulatory procedures for research and product development. The policy which was, in fact, being announced for comment, was a response to both themes.

The two overarching currents of NABC 4—animal welfare and the call for open dialogue—point toward reiteration of a conclusion that was reached at the 1990 NABC meeting in New York State: concerned parties in the food arena have failed to talk with each other, much less communicate. Regrettably, much the same conclusion was reached by the 1992 workshop on Meat and Animal Product Safety. Participants were in strong agreement that differing groups fail to interact and called on scientists to begin to communicate with the public as equal partners.

Defining Food Safety

This failure to interact and communicate became evident in discussions on *how to define safety*. There were those, mostly scientists, responding to the

---

[1] Editor's note: Letters were sent to Vice President Quayle and the heads of HHS, FDA, EPA and USDA expressing NABC's belief that the acceptance of government efforts by the public can be enhanced only when policies are developed and perceived to be developed with appropriate input from all interested parties.

presentations of David Berkowitz, Office of Biotechnology, FDA; Russell Cross, Administrator, USDA/FSIS; and John Frydenlund, Deputy Assistant Secretary, Marketing and Inspection at USDA, who felt that safety should be defined in terms of whether eating a product will cause injury or disease. They urged that scientific principles be used to: 1. assess the probability of foodborne injury to health; and 2. target foods where alterations associated with biotechnology might increase the probability of harm (e.g., allergenicity). A second view was represented by Dianna Hunter, a former small farmer and member of the Minnesota Food Association, who interpreted safety as "feeling confident about one's food." Factors that influence such confidence include whether the food is being produced and provided through a trustworthy source. Many participants agreed that nonscience factors (e.g., social, economic) can influence whether a source is deemed trustworthy and should be considered in the assessment of foodborne risk.

The 1992 meeting illustrated the need for biotechnology industries and high–level government officials to get behind the goal of increasing two–way communication where biotechnologists listen to the nature and shape of public and interest group concerns before formulating their messages about the safety, efficacy and benefits of biotechnology.

Regulatory Policy

A recurring theme in all the workshops was the need for clear regulatory policies for agricultural biotechnology whether for food, pharmaceuticals or animal use or release. In the workshop on Regulatory Issues, participants felt that pharmaceutical products have been foremost in the thinking of regulators who have concluded that the existing framework for biotechnology regulation is adequate and that all forms of regulation should stress product over process. Martin Terry, Vice President for Scientific Activities, Animal Health Institute, expressed the frustration of industry faced with different regulations depending on product classification as a drug or a food. He called the groups attention to both the debate on extra–label drug use in animals and the crisis in drug availability which currently besets animal agriculture. From the environmental perspective Margaret Mellon, Director, National Biotechnology Policy Center, National Wildlife Federation, argued a need for regulatory action, noting that there are several areas, including fish and wildlife, where animal scientists are undertaking biotechnology research in the absence of clear regulatory authority.

The group also discussed how process and point of origin have traditionally been important to consumer acceptance of agricultural products. Virtually every state claims that its soils, climate and farmers produce the best potatoes, onions, wine, pork or something. Furthermore, it was noted that the Food Safety Inspection Service (FSIS) has made decisions based on judgements that are not supported by risk–based reasons. For example, FSIS does not allow lungs to be used in meat products based on the cultural judgement that consumers do not want meat products in which lungs are used—in the U.S. lungs are not classified as food. As such, participants felt that a decision to consider only risk–based regulatory policies for agricultural biotechnology leaves many questions unanswered, for example, labeling and product certification—another continuing theme that emerged in each workshop only to be shelved by each noting the need for a future NABC forum on the issue of labeling of biotechnology food products.

The 1992 meeting explored new issues and revisited several continuing issues. NABC 4 established a series of key understandings that should shape the direction of animal biotechnology research, product development, policy and administration for the coming decade. There was overwhelming consensus that greater public understanding of biotechnology processes and products and greater public participation in the decision–making process was not only desired, but essential, if agricultural biotechnology is, indeed, to be the growth industry of the 21st century.

June Fessenden MacDonald
Deputy Director
NABC

Workshop Highlights and Recommendations

NABC 4 featured four concurrent workshops focusing on areas of concern in agricultural biotechnology: Animal Well–Being, Links of Animal Biotechnology to Human Health, Meat and Animal Product Safety and Regulatory Issues. In the workshops, participants were asked to define and prioritize national issues, reach consensus where possible, and to develop recommendations. The diversity of participants helped to insure that a wide variety of issues were raised; at the same time, diverse values and goals often made consensus difficult. Consequently, workshops were also charged with identifying areas of disagreement both of fact and perception. The following highlights are from reports prepared by workshop co–chairs and reviewed by all participants in those workshops. Any inaccuracies are the result of editing and not the responsibility of the original writers. Full versions of the workshop reports and summaries, as well as background presentations, can be found in Part II starting on page 23.

Biotechnology and Animal Well–Being

The workshop began with a wide–ranging discussion about the concept of well–being. Participants decided that the discussions should be limited to the well–being of animals involved in biotechnology: farm animals and experimental animals. Many felt that new technologies create new problems and raised new questions.

Individual participants listed 15 questions about biotechnology and animal well–being. It was observed that the emergence of biotechnology coincides with greater concerns about animals, increasingly cognitive views of animals, increased distance from agricultural and draft uses of animals, and urbanization and romanticization of animals. Technology is colliding with changing morality. Genetic engineering feeds into these concerns because of the general concern that the manipulation of genes could lead to unnatural beings.

By this point in the discussion the participants were quite polarized. To move the discussion forward it was suggested that some of the fundamental concerns might be identified. Among the 16 listed were: whether it is ever acceptable to utilize animals for human use, whether animal biotechnology poses unique questions about animal well–being and whether biotechnology is qualitatively or quantitatively different from what has come before.

Following an intense discussion on the concerns, individual participants identified some possible harms and benefits to animal well–being that arise in the context of biotechnology. A highly unrepresentative straw poll was then done in order to see which of these possible harms and benefits the participants most wanted to discuss. The four possible harms (1. diverting resources away from improving traditional husbandry practices; 2. loss of genetic diversity; 3. proliferation of genetically defective animals who suffer disease as models; and 4. thinking of domestic animals as human artifacts), and the four possible benefits (1. removal of genetic defects from animal populations more rapidly; 2. better understanding of animal well–being; 3. permitting increased disease resistance; and 4. more efficient production leading to the use of fewer animals) receiving the most support, along with the possibility that animal biotechnology may lead to healthier products for both humans and animals, formed the basis of much of the remaining discussion.

CONSENSUS STATEMENTS

Weighing the broad spectrum of issues related to biotechnology and animal well–being, participants were able to reach agreement on four consensus statements:

1. Biotechnology may contribute to animal well–being, but it is not the only approach to improving animal well–being.

2. There should be responsible, systematic investigation of the benefits and harms to animals that may be associated with biotechnology.

3. It is acceptable under some conditions to use animals for human use.

4. Animal biotechnology has the potential to contribute to the "three Rs" in animal experimentation: reduction, refinement and replacement.

RECOMMENDATIONS

1. With respect to animal well–being, criteria should be developed for responsible research and application of specific biotechnologies in animals. The full spectrum of opinion should be represented in the development of these criteria. These criteria should be periodically reconsidered in the light of changing circumstances.

2. The benefits and harms noted should be taken into account in developing these criteria.

3. Animal biotechnology should not be used in ways that impose great costs in animal well–being while achieving only minor human or animal benefits. When there is the likelihood that a procedure will cause great suffering to animals, alternatives should be sought.

Links of Animal Biotechnology to Human Health

As a result of the advances made in molecular biology over the last decade, the fields of animal agriculture and human medicine have come to share a wide range of techniques and models. These profound changes in the research process have raised a series of issues with respect to the use of both farm and traditional laboratory animals in research. The discussions in this workshop focused on these issues.

HUMAN HEALTH CONSIDERATIONS DRIVING AGRICULTURAL RESEARCH

In recent years, public concern about food safety and nutrition has played an increasing role in animal agricultural research. The public is also concerned about the disclosure of the contents of food and food products as well as about broader marketing issues (e.g., product claims).

In addition, new biotechnologies blur the lines between nutrition and pharmacy, making possible the creation of what have been variously called "nutraceuticals" and "pharmafoods." These products, often of animal origin, serve a combination of nutritive and therapeutic goals. They raise complex issues of regulation, food safety and consumer education.

ETHICAL USE OF ANIMALS

Some argue that the use of animals as food or in research is itself unethical. Others argue that humane treatment of animals is the major concern. The workshop participants agreed that it was not clear just what is ethical. Moreover, they were concerned with methods used to accommodate the wide range of views on the subject found in our diverse society. They also questioned whether current guidelines on the use of animals in research, often written before the advent of the new technologies, are adequate morally.

ANIMALS FOR BIOLOGICS AND THERAPEUTICS

The use of animals for the production of vaccines and therapeutics has a long history. Workshop members indicated that the widespread use of animals as living "bioreactors" to produce chemicals of value to humans differed from other uses of animals, (e.g., in food and fiber production). They expressed concern as to what, if any, ethical implications were associated with it. Moreover, the use of animals as bioreactors raises some practical questions. For example, there is the problem of what to do with the carcasses of these animals. Should they be allowed to enter the food chain?

Animal bioreactors also pose problems of containment, welfare and management, raising the question of whether it would be more desirable to have certain species earmarked for this purpose and not used for food. This, in turn, raised the issue of whether whole animals or cell cultures should be used for screening of therapeutic products.

SOCIETAL CONTEXT OF SCIENCE SHARED BY AGRICULTURE AND MEDICINE

Research rarely takes place outside a larger social context. That context provides both the limits and opportunities for research. A central issue in this workshop was how (or whether) to integrate private and public research at the agriculture–medicine interface. Another key issue was the distributive aspects of this type of research.

The group also acknowledged that new linkages between the medical and agricultural sciences will be influenced by the current state of food, agricultural and medical policy. At the same time, the discoveries and inventions stemming from this research will have a considerable impact on food, agricultural, and medical policies. In addition, workshop participants wondered whether the current institutional structures (especially at universities) were adequate for the new linkages between agriculture and medicine.

Finally, there was a general consensus that greater public participation in the decision–making process was both necessary and desirable.

RECOMMENDATIONS

1. Stronger links need to be developed between agricultural and medical research relating to biotechnology. *Among mechanisms to do so are centers, incentives for joint programs, funding, etc. This will require further integration and institutionalization of joint agricultural and medical programs. Such linkages will need to include an examination of the ethical, economic, social, institutional, and legal ramifications of these changes.*

2. More resources from molecular biology should be devoted to genome and other research in an attempt to ultimately spare animals from direct use in research. *It should be thereby possible to shift largely from whole animal to organ, tissue or cellular systems.*

3. Explore the moral implications of the use of animals in medical and agricultural research. *Issues in the area are currently inadequately examined, and thus, there is not yet an adequate moral framework for making decisions about this type of research.*

4. Provide for education of and dialogue among all the participants in the debate.

5. Improve the agenda–setting process that insures that resources are properly allocated and that all interested parties are involved in the allocation process.

6. Improve the guidelines to aid in determining appropriate circumstances for patenting animals, tissues, and cell lines.

Meat and Animal Product Safety

Workshop participants identified some potential safety problems for discussion. These included unanswered questions about bovine somatotropin (BST), allergenicity and questions about a number of products for which there are, as yet, no data bases. Participants also discussed the promise for new biotechnologies to produce diagnostic tools for food safety testing of animal products.

Finding common ground was more difficult and frustrating once the group moved past the fairly narrow, but controllable technical hazards to the myriad of intellectual and social elements that people bring to a decision about the safety of any entity, food included. At this point, participants stepped back to list the major concern of each of the participants about the safety of biotechnologically produced meat and animal products. The items fell into four different areas. Small groups were formed to discuss these issues and bring recommendations back to the total workshop group for discussion.

THE SAFETY OF TRANSGENIC ANIMALS AND ANIMALS ADMINISTERED RECOMBINANT DNA PRODUCTS

In the area of use of transgenic animals to produce pharmaceutical agents for use by humans, the major safety concern was that these "pharm" animals may enter the human food supply, but before they do, their safety must be assured.

1. ***All workshop participants agreed to the need for a data base on the nutrient composition and levels of relevant hormones and residues in these animals to reassure scientists and the public that there are no detectable differences from levels of these substances in traditional animal products.*** There was not consensus in the group as to how extensive the data base would be and what it would contain.

In the area of animals administered recombinant DNA products: 1. hormones; 2. vaccines; and 3. direct–fed microbials; there was consensus that the regulations under the National Environmental Protection Act (NEPA) and the testing protocols for vaccines were probably adequate. FDA has the authority to regulate direct–fed microbials, but the group felt it has not been doing so.

2. ***FDA should investigate direct–fed microbials more carefully in the future when applications for recombinant products are received.***

Another concern expressed was about long–term consequences of breeding transgenic animals. The concern here is the unknown potential for unexpressed genes to cause other changes in animals that may not be expressed for several generations.

3. ***The final recommendation in this area speaks to the need for remaining aware of the possibility of cloning defects in embryo transfer and cloning experiments.***

BIOTECHNOLOGICAL TOOLS TO ENHANCE FOOD SAFETY AND QUALITY

4. Recognizing that animal products are the major source of microbial contamination in the food supply, the use of DNA probe assays and immunoassays for the detection of pathogens is to be strongly encouraged.

Biotechnology is the most promising source of tools that can yield rapid, sensitive, specific and cost–effective diagnostic tests for the presence of microbiological pathogens, antigens, toxins and other compounds of interest to improve food safety. New diagnostic capabilities can also be used to detect adulterated foods and as a screening method for allergens in the food supply. The group also discussed how genetic markers offer the potential to improve the healthfulness and safety of the food supply.

5. Research and application of these tools should move ahead rapidly. They endorse continued research on the use of the genetic makers techniques.

DEFINING FOOD SAFETY

Some participants argued the present definition of "safe," relative to foods, is too narrow, ignoring quality issues as well as the fact that food safety is a social construct. They felt that social, economic and political issues should be evaluated concurrently with the evaluation of efficacy and human and animal safety. Others disagreed with all of these ideas and argued for maintaining the present system of relying solely on technical data for safety decisions. The latter participants did recognize that social, economic and political issues should be discussed, but there was no agreement about whether the mechanism should be separate from, or integral to, the present system.

6. The larger issue here is how to define food safety.

COMMUNICATING WITH THE PUBLIC

This section of the report and recommendations are premised on a consensus agreement that the public has a stake in maintaining public institutions provided they are responsive to public needs. Many (but not all) scientists have perceptions and biases that are quite different from the various perceptions and biases of public groups which makes it difficult for scientists to be good communicators. There is also the serious problem of lack of support for these activities in the reward structures of institutions and of an imbalance in funding going to high technology research versus research in policy and communications. These were all considered in the following set of recommendations which were endorsed by all workshop participants.

7. There is a body of knowledge about communications that scientists should use to improve the dialogue with the public.

8. Regional research projects should be promoted and funded, and the National Research Initiative should be encouraged to put more funding into its policy and marketing line item to promote public understanding of agricultural biotechnology.

9. Interdisciplinary work between the biological and social sciences should be promoted and recognized as critical if serious progress in this area is expected.

10. In all grant proposals, the technical significance and relevance of research should be communicated in terms the general public (or anyone outside the particular discipline) can understand.

11. Continuing education programs should be developed for scientists to teach them how to more effectively facilitate two–way communication between scientists and the general public.

12. The public, starting at the elementary school level, would be well served by educational programs on the social, moral, economic, political and scientific issues surrounding biotechnology.

In order to accomplish any wide–ranging change in faculty behavior in these areas the group suggested that it will be necessary to re–envision the mission of the land–grant colleges to serve all their publics and recognize that the responsibility for this is shared by all institutions of higher education. This will change the weight given to public service or extension activities in promotion decisions and bring this area into better balance with research and teaching.

Regulatory Issues

The charge to the workshop participants was to identify and examine issues arising in regulatory treatment of animal biotechnology. Free–ranging discussion among individuals with different perspectives followed. While consensus was not sought nor achieved on the specific issues identified, these issues were deemed worthy of consideration by one, some, or many of the members of the group, and as such help to illustrate the range of concerns in the regulatory arena. Common themes of agreement did emerge and these were captured in the form of four issue statements or recommendations at the end.

The following issues and gaps have been identified in the regulatory process:

—At the research stage, there are no mandated guidelines/regulations for industrial research of animal biotechnology.

—In field testing, there are no regulations for release of fish, wildlife, insects or pets; for micro-organisms in livestock feeds or for zoonotic pathogens of animals and humans. Also there is no mechanism to deregulate similar genetically modified organisms that have been proven to be safe based upon previous case studies;

—Implementation of the ABRAC developed guidelines should govern agricultural research in the area of field testing;

—There is an inability to gain access to some information on health and safety of products because of "confidential business information" designation.

—In the food safety area there are gaps that are currently not covered by any regulations including disposition of transgenic animals and whether fish, seafood and wildlife should be included.

—In efficacy testing, the issue of whether transgenic animals used as pharmacoreactors should receive special attention from FDA was raised.

Recognizing there should be representation of broad interest, the public's role in regulatory debate was discussed. Possible mechanisms identified for improved public access included: 1. legislation regarding public participation in regulating decisions across the board; 2. publication beyond the *Federal Register*; 3. improved representation in decision–making processes; 4. open forums; 5. research on opening up scientific decision–making process; and 6. rebuilding public trust and regulatory transparency.

Other issues considered were the role of states and industry in the debates, public education, and communication. Consideration was given by the group to the level of information available for consumer choice.

Workshop participants also considered technically based regulations vs social/ethical/economic impact considerations. They noted that regulations can impact not only in the U.S., but also on international trade, as well as trade relationships with Third World countries.

RECOMMENDATIONS

*1. **The regulatory gaps delineated deserve serious investigation.** NABC may wish to establish a committee or other mechanism to assist this investigation.*

*2. **A more acceptable policy–making process for rules of broad applicability would be clearly understood or known (not ad hoc), transparent, and participatory.** The group viewed the process by the Council on Competitiveness in the Office of the Vice President, leading to the May 26, 1992 FDA food safety decision, as falling short of the goals for an acceptable process.* (Editor's note: At the final plenary session, a recommendation was made by those present (no opposition was voiced) that NABC respond urging future processes of policy development be open and include all interested stakeholders. NABC sent letters to Vice President Dan Quayle and the heads of HHS, FDA, EPA and USDA.)

*3. **Social, economic and ethical questions need to be explored.** What role do/should these issues have in research, development and approval processes for commercial use of new products? When should these factors be considered, relative to, but not necessarily as a part of, the regulatory process?*

4. With broader representation, (such as food processors and consumer groups), NABC should conduct further exploration of the relationship between the government's regulatory role, particularly the safety statutes and issues of choice, such as labeling provisions.

Bill R. Baumgardt
Director, Agricultural Experiment Station
Purdue University
Member, NABC Council

NABC 4: Opportunities and Challenges

We, the participants in NABC 4, have been presented with a tremendous amount of information—facts, perceptions, views, values and impacts on animal biotechnology. It has been a stimulating, mind-stretching and, yes at times, a stressful experience. In this meeting, we have not all really "heard" the same things even when we were in the same sessions listening to the same speakers Certainly, we have developed individual impressions. We leave this meeting with many of us feeling a little uncomfortable—business representatives feeling as if they were at times considered the "bad guys." Ditto for government representatives and scientists, animal rights proponents and those with environmental interests. Let me suggest that NABC must be doing something right when participants *do* feel a bit uncomfortable, a little bit less sure after actually listening and hearing views different from their own; when representatives of different groups have had to articulate their positions to fellow participants who are unfamiliar with, and even disapprove of, their positions and actions. This forces communication, perhaps for the first time, *with* one another rather that *at* one another. If you are feeling just a bit less sure, if you have a bit more insight to the perspective of a different group, if you now know someone "from the opposition," I declare NABC 4 a success.

Let me just review with you some of the opportunities and challenges we encountered as participants in this meeting. First, we need to examine some of the "specs" of the meeting that we are pull together here. As a benchmark, NABC Chair Ralph W. F. Hardy, in his charge to the meeting made several key points: 1. We are living in an era of biology with rapid change; 2. Yesterday's science is perhaps but dreams; 3. Today's science represents possibilities; 4. Tomorrow's science will bring realities; 5. Dr. Hardy implored us to recognize that "risk" is *product*–based, not *process*–based; 6. Furthermore, he pointed out that we now have the knowledge base to allow a thorough look at risk; and 7. Finally, Dr. Hardy reminded us that the mission of NABC is that of "providing an open forum for exploring issues in agricultural biotechnology." In conducting that open forum we come together to speak, to listen, to learn and to participate in meaningful dialogue and evaluation of potential impacts of agricultural biotechnology. As I reflect on how well we responded to the charge, I note both some rewarding opportunities we have taken and others we left untouched as well as several challenges we, as individual participants, will take from this meeting.

OPPORTUNITIES

With this assemblage of diverse views and interests, I can only wish we had had more time to discuss more of the opportunities placed before us. Let me share my views about a few opportunities we missed.

The issue of patenting came up several times. As keynote speaker Dorothy Nelkin put it, "Patenting of animals has become a lightning rod." I was disappointed that we did not take the opportunity to dissect the patenting issue and to examine the components. (I realize there is a special optional seminar that will place additional emphasis on patenting, but not all participants will be at that activity.[1] There is a general perception that patents (in any field) inherently mean: 1. secrecy and 2. making money. I have learned, after having spoken to various university patent officers, that they support my own experiences which suggests the following: 1. Most patents do *not* return many dollars in royalties; 2. The patent process makes information known to the public. One of the requirements for issuing a patent is that the details be disclosed; 3. The main reason why universities patent things is to encourage commercialization. Most products coming out of university research that may be patentable still require further research and development before becoming available to users. That means that some business organization must make additional investment of likelihood of the business getting a financial return on their own funds to bring the patented discovery to commercialization. There usually is significant risk about the investment; 4. It seems to me that licensing is the key issue that really should be addressed under the protection of intellectual property by a university or other public sector entity. If a university owns the patent they can control the release for the best benefit to those who should be receiving the benefit. In many cases, patents are licensed by the public sector to the private sector on a nonexclusive basis. However, sometimes it may be necessary to provide an exclusive license so that a company can have sufficient incentive to cover the additional costs to complete the research and development and then test the marketability of the product or idea. There are times when an exclusive license may be the most appropriate way to proceed.

Another point about licensing is that the intellectual property can be licensed for less than the life of a patent. Thus, even if a university decided that it was in the best public interest to give an exclusive license, that would not have to be done for the life of the patent. It could be made available exclusively for a particular company's use for, let us say, five years and after that then, the university may be in a position to license it to other companies. This sort of arrangement provides the initial company the lead time to complete the necessary development and have some headstart in the marketing area as an incentive for potential return on that upfront investment. If the university has not patented intellectual property in the first place, they would not have the option to control the route to commercialization.

I also noted that Dorothy Nelkin made an observation of a general nature: biotechnology is not the cause of the decline of family farms and banning of patents would not likely reverse the trend in decline in the number of family farms. It was a

[1] Editor's note: *NABC Occasional Paper #1, Ethics and Patenting of Transgenic Animals*, is available from the NABC office.

challenge that was not picked up by any other speaker, or to any significance, in any workshop.

NABC 4 provided an opportunity for fairly rigorous examination of animal biotechnology implications. I was surprised that there was not (at least in the sessions I attended) more examination and analysis of issues (or the lack of) related specifically to each of the different tools and uses of biotechnology for animals: 1. Of course, artificial insemination has been common for 40 years and the subsequent advent of frozen semen represents really an early "biotechnology" that has had tremendous positive impact on the dairy industry and on the efficiency of dairy food production. At the present time, nearly three–fourths of the dairy cows in the U.S. are artificially inseminated using enhanced genetic material; 2. Control of reproduction involving ovulation, super-ovulation and estrous cycle regulation; 3. *In vitro* fertilization, perhaps with spermatozoan carrying X or Y chromosomes thus enabling control of sex of the offspring; 4. Embryo manipulations including divisions (cloning) and nuclear transfer; 5. Production of transgenic animals. Although this area of animal biotechnology perhaps attracted the most attention, it is, in fact, currently only a very small component of all uses of biotechnology for animal agriculture; 6. Marker–assisted genetic selection through the use of techniques for gene mapping. This technology may turn out to be the most powerful tool for the intermediate time range; 7. Vaccines produced through biotechnologies; 8. Diagnostic tests; and 9. DNA probes. Each of these has a different set of advantages and issues. It is unfortunate that we did not take this opportunity to look at each of the technologies individually, providing an analysis of the issues and potential outcomes of various alternatives.

OPPORTUNITIES TAKEN

Next, I would like to examine some of the opportunities taken, or overarching themes, which appear to me to at least link, if not tie together, the participants of NABC 4.

The first I would identify relates to animal well–being. Workshop participants were hard pressed to come up with a precise definition that was acceptable to all. While no attempt was made to achieve consensus overall, it seems to me that many participants in the workshop were saying: 1. Pain and suffering are no longer the only criteria to be considered; 2. Some feel "animal welfare" may be a minimum and that "well–being" means something above that in terms of production, reproductive performance, etc; 3. Animal well-being is important; and 4. Animal well–being is an important component of the biotechnology dialogue.

A second theme relates to meat and animal product safety. Again, I heard many people saying that healthy transgenic food animals are likely to be just as safe as a source of food as that from "traditional animals" from which they are derived. That goes along with the concept of looking at "product" and not "process." I also heard many saying they wanted to know when their food was biotechnologically produced—a concern about labeling. A related theme is the clear need to continually examine, and perhaps broaden, our definitions and agendas for issues such as food safety, especially where food allergies are a concern.

Finally, I often heard (and yet make no claim that there was an effort to achieve consensus) that biotechnology is not inherently, or universally, bad.

CHALLENGES

In addition, there were a few challenges to us as participants. These remind us that we all need to continue to listen, learn and share. One challenge to NABC was the need to broaden the representation of groups in the audience. It was pointed out that we especially need in the future to encourage greater participation of farmers, of food processors and, certainly, of consumers (as one individual put it, of "eaters").

A second challenge: the regulatory process continues to be a concern and one for which it is very difficult to meet even an approximation of consensus among all of us and the groups we represent here at this meeting. Clearly we need more dialogue and meaningful interaction all through the year, not just for a few day once a year.

Perhaps the greatest challenge running through sessions and workshops even approaching consensus is two–fold: 1. the public will be more involved in decisions about biotechnology whether or not the traditional decision–makers invite them; and 2. there is a critical need for enhanced true communication. In discussing the public, we were reminded that we should not view public participation as an impediment to change. This meeting reinforces some observations I have made from other vantage points, namely that "the public" is showing distrust, confusion and frustration not just about biotechnology and science, but also with government, legislators, etc.

I believe *all* of us must be seriously concerned that science may continue to lose credibility in the eyes of the public. Because science holds many keys to the future, a critical question is, "How are we going to develop and use science (e.g., biotechnology) for the public good?" Scientists have a responsibility to provide information about science in understandable ways for all of society. "The public" must be invited into the decision–making process as full, and informed, participants.

The second part of the major challenge from this NABC meeting is the need for enhanced communication. Participants clearly declared that by communication we do not mean "we need to educate the public." Some of the sub–points in those discussions, and which I would challenge you to think about, are as follows:

—We must keep talking about issues and be open to compromise.

—Scientists have *not* done a good job of communicating what is going on. There is a level of mistrust.

—Real communication is not just reaching someone with *your* message; it is interaction among equals who may not necessarily agree. As someone put it "don't talk only with those who agree with you ... you'll never learn anything."

—Consumers are removed from agriculture and the production of food and fiber. Since they do not see the various steps in the process they then have difficulty understanding this new animal biotechnology.

—The science community cannot just say that public concerns are unfounded. Public perception drives policy.

—And finally, special attention is needed to work with the media.

I cannot leave this topic of communication without pointing out an old axiom, but one worth repeating—*"we" must not just talk to our "own kind."* That is true whether our own kind means scientists, administrators, regulators, advocate and consumer group representatives, or any other segment of society. That is what NABC is all about—not talking only to our "own kind."

I suggest that each of us go back from this meeting to our respective organizations with some action items. One that strikes me very clearly is about the development of public policy. I am reminded that it is said about real estate that three things are important—location, location, location. If science is to have a role in the development of public policy it seems to me that three things are important—*communication, communication, communication.*

I would hope that all of us, regardless of what group we consider ourselves to be affiliated with, would seriously consider these steps in communication: 1. Listen to a broad segment of society; 2. Listen carefully; 3. Frame a response; 4. Go back to our audiences with patience and persistence; 5. Listen to the audience once again; and 6. Keep up the cycle. It must become a part of our regular way of doing business.

Throughout this meeting we also heard the need for linkages—for coalitions. Each of us has a responsibility for our own organization in terms of seeking linkages and coordination. NABC provides a unique opportunity to enhance those efforts for appropriate development and use of agricultural biotechnology.

CONCLUSION

In conclusion, let me confess that I have not been able to tie the box representing this meeting closed. In fact, I have decided that we do not have a box at all, but rather a vehicle. I urge each of you to consider the opportunities and challenges that we have shared during this meeting as fuel. It is now up to us to make that vehicle move forward. Let me leave you with the urgent plea that you communicate openly and fully with all stakeholders in agricultural biotechnology.

On a light note about a serious issue, perhaps it is not biotechnology per se that causes things to change. I believe that on his late night show Jay Leno said recently "the new McLean burger shows how times have changed. Now you find water and sea weed at McDonald's and oil and grease in the ocean."

Dr. W. Edwards Deming, (the American who finally gained fame for his success in rebuilding Japan's industry following World War II) repeatedly stresses the importance to "optimize the system so everybody wins. The best solution is for everybody to win." NABC provides an open forum, a level playing field. Let us all work to keep the field level and the vehicle, fueled by this open forum, moving.

Finally a quote by Ralph Waldo Emerson, the American scholar, "This time like all times is a very good time, if we but know what to do with it." I challenge us to communicate with each other and all the public. That is what we should do with "It" (our time).

PART II ANIMAL BIOTECHNOLOGY AND THE PUBLIC GOOD

Charles W. Stenholm
Member, U.S. House of Representatives, Texas;
Chair, House Agriculture Subcommittee
on Livestock, Dairy and Poultry

*Daniel B. Waggoner**
Visiting Fulbright Scholar
Massey University, New Zealand

Public Policy and Animal Biotechnology in the 1990s: Challenges and Opportunities†

American agriculture has enjoyed momentous success over the past 50 years as measured by the quantity, quality, variety and cost of food and fiber. In the years proceeding World War II, the U.S. agricultural sector has experienced a significantly high rate of growth in productivity—a level more than three times the magnitude of the nonfarm industrial sector.

Into this agricultural system with all its strengths, complexities and challenges—both biological and political—comes biotechnology. Because of its importance to increased competitiveness in today's expanding global economy, biotechnology is viewed as one of the keys to U.S. agriculture's continued success in the years ahead. Moreover, it is predicted that the world's population will increase at a rate of approximately 90 million people annually. At this rate, the current global population of just over five billion is expected to double during the next century. World hunger and malnutrition will not be simply problems of inequitable distribution. Expanded food production will be essential to accommodate the nutritional needs of this rapidly growing global population. Herein lies one of biotechnology's most pressing demands.

AN INDUSTRY COMES OF AGE

After years of speculation and commitment, agricultural biotechnology is moving slowly from the research laboratory to the barn, the field and the processing plant. The puissance of agro–food biotechnology is no longer fantasy. Already, diagnosis of disease using biotechnology tools is a reality which is

* Former Staff Director, U.S. House Agriculture Subcommittee on Livestock, Dairy and Poultry.

† The views and opinions expressed in this article are the authors and do not necessarily represent the official policy or interpretations of the U.S. House Committee on Agriculture.

changing the face of both human and animal medicine. We are now in what some call the age of biology, moving from the age of chemistry.

The U.S. has maintained its preeminence in biotechnology, bolstered by strong research programs and well–established foundations in pharmaceuticals and agricultural science. For instance, in 1991, sales from biotechnologies totaled approximately $5.8 billion, an 18 percent increase over 1990, with net exports exceeding $600 million (Burrill and Lee, 1991; Raines, 1991). Furthermore, the Council on Competitiveness in the Office of the Vice President (1991) projects that by the year 2000, biotechnology will be a $50 billion industry. Currently, private industry spends approximately $2.1 billion annually on technology development (Office of Technology Assessment, 1991a). The federal–state agricultural research system spends roughly $1.9 billion annually on agro–food biotechnology research and development (Office of Technology Assessment, 1991c).

These figures merely underscore the fact that what scientists have come to understand thus far about plants and animals is impressive. Moreover, this basic knowledge has been rapidly carried forward by a whole host of viable applications.

IMPACT OF BIOTECHNOLOGY ON ANIMAL AGRICULTURE

Over the next 15 years, American farmers and ranchers will be offered an extensive array of new technologies that could revolutionize food animal production. Ongoing research in the areas of computers, information systems and processing, robotics, controlled environments and biotechnology are expected to provide numerous on–farm practical applications (National Research Council, 1990). Such technologies point to more efficient growth rates, less feed per unit of output, improved disease resistance and increased prolificacy (Van der Wal et al., 1991).

Today, biotechnology has provided animal agriculture with safer, more efficacious vaccines against viral and bacterial diseases such as pseudorabies, enteric colibacillosis, and foot–and–mouth disease. We are beginning to seek answers to questions regarding complex systems that only a few years ago we could not even think to ask. This increased ability is particularly important in light of the fact that food animal products account for approximately one–half of all U.S. agricultural revenues on an annual basis. Producing leaner, high quality meat and meat products to satisfy today's health–conscience consumer is of paramount importance (Pearson and Dutson, 1990; Kopchick, 1992).

Further, the first few commercialized on–farm animal biotechnology products will be of particular significance since: 1. they will heavily influence public attitudes about other emerging products and applications; 2. will establish substantive and procedural precedents in the legislative and regulatory arenas; and 3. will impact the future willingness of the corporate community to invest in like or similar product research and development (Kalter, 1985; National Research Council, 1987; Office of Technology Assessment, 1991d).

Unfortunately, there is a lack of recognition in some circles that biotechnological applications complement, rather than replace, the traditional methods used to enhance agricultural productivity. In reality, many of the so–called "new" biotechnologies involve concepts based on centuries–old applications (Moses and Cape, 1991). Bovine somatotropin (BST) is an interesting example—a product that elaborates familiar disciplines such as breeding, animal nutrition, animal physiology and veterinary science, supplemented with the basic disciplines of molecular genetics, biochemistry, microbiology and bioprocess engineering (Office of Technology Assessment, 1991d).

CURRENT CHALLENGES FACING ANIMAL AGRICULTURE

Meeting the challenges of international competitiveness, sustaining a high quality food supply, preserving natural resources and protecting the environment will require a heightened level of knowledge over and above what was required to solve previous problems of years past (National Research Council, 1989). In fact, an array of thought–provoking questions are being posed from both within and outside the agricultural sector. For example, how can the safety of biotechnologies, which may be used in food production or processing, be systematically evaluated? Will the release of genetically modified organisms into the environment pose threats to human health or to natural ecosystems? Is new legislation necessary to regulate the agro–food and fiber products that are likely to be developed utilizing biotechnologies? These and other intricate questions are being voiced with a heightened urgency.

Clearly, the issues and strategies have become increasingly complex. Legislative authority and jurisdiction have become widely dispersed among several congressional committees and subcommittees with differing and like perspectives. Nonetheless, in the end, an effective biotechnology policy must knit several dimensions into a coherent framework—including basic research, development and application, marketing and economic competitiveness, effective regulation, ethics and public policy (Office of Technology Assessment, 1991a).

OVERCOMING OBSTACLES TO CHANGE

Worldwide, biotechnology is debated on three fundamental planes. The risks and benefits are disputed on scientific grounds, socioeconomic grounds and on the basis of public perception (von Oehsen, 1988; Wald, 1992). As with any other new technology, many questions of adjustment to change are posed. Concern about the effects of technological change has been a constant in the history of industrial development. But how should a democratic society establish public policies on advanced technical issues like biotechnology?

The primary difficulties inhibiting adoption would appear to lie in the provinces of administration, economics, management—and politics. The political debate surrounding biotechnology begins at the edge of scientific knowledge and lies in the realm of "What if?" Consequently, the barriers are both technical and institutional (Ruttan, 1991). Although we live in a society where the words of acknowledged experts are often received as gospel, our fascination with authority shows some indication of waning. Scientists, industrialists, politicians and educators have been found to be as fallible as other human beings and their "expert" information is greeted with scepticism by some, and with open defiance by others.

Today, the agricultural research community, and production agriculture in general, face several formidable outside forces. Such influences are frequently referred to as *externalities* which can either have a positive or negative effect upon agricultural research and its use—particularly for agro–food biotechnology. The ultimate judge of emerging technologies will be the consumer—whether that be the farmer, homemaker or general public (Harlander, 1991). It is they who will appraise the merits of a particular product or process and determine its success or failure.

UNDERSTANDING CONSUMER CONCERNS

Within the past few years the popular press has captured the public's attention with the perceived role biotechnology might play in agriculture, citing both positive and negative aspects, whether realistic or wildly speculative. Further, many of the terms used in current discussions of biotechnology have negative overtones. For example, words such as "genetic manipulation" and "genetic engineering" have a pernicious ring to the general public, and it is significant that those campaigning against BST continually refer to the product as "bovine growth hormone."

Part of the problem surrounding the broad acceptance of biotechnology stems from the frequently espoused concern that the processes of scientific research, and the applications derived therefrom, seem difficult to access and thus opaque, especially to the ordinary citizen. An examination of the testimony received by the U.S. House Committee on Agriculture over the last several years would seem to indicate that concerns regarding biotechnology and other advancements facing animal agriculture fall into two broad categories—those relating to animal and consumer safety issues and those relating to social and ethical issues. Furthermore, the testimony submitted by various public interest groups can be summarized under four question headings: 1. Is it natural? 2. Is it right? 3. Is it fair? and 4. Do we need it?

Many characteristics have been identified in the literature that appear to influence consumer acceptance of innovation. Among them are relative advantage, compatibility, complexity, trialability, observability—and risk (Herbig and Day, 1992).

Undoubtedly, citizens in the U.S. and around the world are going through an often mind–numbing debate about risk and reward in many aspects of their lives. Whether it is food safety, car safety, atomic energy, liability laws or even nuclear weapon prolification, we are in the midst of a debate about the "balance" of risk and reward in society. Public attitudes can vary, often for reasons beyond the influence of more or better information about potential personal, environmental or socioeconomic consequences of a technology. Moreover, some people may react negatively to the perceived impacts associated with anything new or innovative.

What happens if the consumer does not change and refuses to use a technology? Should one blindly accept the scientists' opinion of what is best? Do the vested interests which exist in a company or industry for an innovation mean that consumers must accept their decision? There are a whole host of questions and public concerns which must be properly considered and adequately addressed if we are to clearly see what the perspectives are for introducing biotechnology in farm animal production. The creation and maintenance of the public trust is surely one of the pivotal tasks to be undertaken (Harlander, 1991; Stenholm and Waggoner, 1992).

THE NEED TO COMMUNICATE

There are those who ask, "Why do we need to understand the consumer's acceptance mechanism?" The answer is simple and straightforward: If scientific advances are to be allowed to provide an affordable, nutritious and sustaining diet for all, the information gap between science and the lay public must be narrowed and the consumer's perspective understood. Once the public is knowledgeable and properly informed, the word "biotechnology" in connection with food production should not raise a red flag of fear, but rather present thoughts of reduced food costs, more nutritious food supplies, a safer food supply and a healthier environment in which to raise one's family.

In an effort to foster the public trust, greater efforts are needed in providing useful information about the working areas of biotechnology and its applications in animal production (Office of Technology Assessment, 1991d; Moses and Cape, 1991). Such information could help support an open and balanced public debate, and thus, form a firmer basis for sound decision–making and sufficient monitoring. We must ensure that our systems of oversight, legislative and regulatory, are transparent and open to full participation by all responsible parties (Stenholm and Waggoner, 1992). At a time when more and more of American life is rooted in science and technology and when the nation's economic well–being depends as never before on its understanding and utilization, the federal government cannot be complacent about the public's interest and confidence in science. Of course, there will always be some degree of risk, but as understanding grows the circle of consensus will widen.

A sure prescription for disaster is for each of the many sides of the discussion to treat the others with contempt. Sincere understanding of the obstructions, and a mutual willingness to confront them, is a critical first step toward positive conflict resolution.

FLEXIBILITY IS THE KEY

A competitive and profitable agriculture will depend on flexibility—on the industry's ability to respond to, and operate within, an uncertain and rapidly changing environment. This means we must learn to view agriculture as a system. We cannot be "smart in the parts" and "dumb in the whole."

Today's current federal meat inspection scheme is a useful example where flexibility and science–driven decisions are presently in short supply. After visiting numerous modern meat processing facilities operating all across America, one should recognize that meat hygiene is a complex subject involving aspects of animal husbandry and physiology as well as food technology and microbiology. Growing scientific consensus supports the view that the allocation of inspection resources in modern meat and poultry production and processing enterprises should reflect a distribution according to risk rather than a distribution according to the classical rules of meat inspection which rely heavily upon human organoleptic methods of detection (General Accounting Office, 1992). Unfortunately, resistance from the inspector's labor union and ongoing concerns within consumer advocacy groups has so far prevented the full implementation of a truly science–based, risk–oriented scheme.

In protecting the public health, a stable and sound regulatory regime is essential. However, since the agro–food industries experience rapid break–throughs in the discovery of new techniques and products, it is important to ensure that regulatory systems do not lag behind emerging, proven developments (Office of Technology Assessment, 1988; Council on Competitiveness, 1991).

The key to improved competitiveness will lie in a relatively more flexible industrial structure and social organization capable of quickly taking advantage of new technological advances (Office of Technology Assessment, 1991b). Further, industry should not rely on the regulatory process as the only mechanism to influence public opinion. The biotechnology and food industries need to strengthen and promote their own credibility to reduce the burden on, and necessity of, the review and inspection processes.

THE INTERNATIONAL DIMENSION

Biotechnology knows no international boundaries. A number of nations have targeted biotechnology as being critical for future economic growth—giving rise to several nationally based research and development programs (Office of Technology Assessment, 1991a). As a result, agricultural systems throughout the world continue to adopt new and advanced technologies that enable them to become more efficient and competitive in developing new markets and capturing old markets for their agricultural products (National Research Council, 1987). The Japanese government, in particular, has organized research consortia among companies, has sponsored research into biotechnology by industry, and has greatly enlarged its overall funding of biotechnology research (Yuan and Dibner, 1990). Consequently, the speed in which innovations are adapted to commercial purposes is a critical factor in achieving and maintaining America's own international competitiveness.

In many critical, high–technology sectors such as biotechnology, American firms are facing competitors whose business risks are shared by their governments (Office of Technology Assessment, 1991b). The U.S. approach to promoting particular industries has been one of refrain and "hands–off," the underlying belief being that the national economic interest is best served by free and fair competition in the marketplace—at home and abroad. Does this approach still make sense in a world where governments in most advanced industrial nations, including those of our most able competitors, are cooperating with private business to promote critically important industries? Maintaining the productivity and competitiveness of U.S. agriculture in the public interest requires a proper balance between public and private sector support for technological change. To move agriculture toward new market opportunities, government must not only support worthy research endeavors, it must also be a partner with industry in moving promising ideas and applications from the lab to the farm (National Research Council, 1989). Further, in such a research–intensive industry, the need to protect innovation is crucial. Many researchers and industry leaders cite protection of intellectual property as being of paramount importance to preserving competitiveness in biotechnology (Office of Technology Assessment, 1991a).

Patenting, licensing and regulatory issues are all areas that affect the rate and cost of technology transfer. They play necessary roles in advancing technology transfer and facilitating the commercialization of research results, especially in capital–intensive fields such as biotechnology. Consequently, efforts should be continued to harmonize and improve intellectual property protection procedures throughout the world (Office of Technology Assessment, 1989; Council on Competitiveness, 1991).

MARKET STRATEGY ESSENTIALS

Will the United States retain its preeminence in biotechnology or will products and services derived from biotechnology be more successfully commercialized in other nations? Acceptance of a new agricultural product seems deceptively simple. Our most superficial experiences tell us that good ideas should work and fittingly render a tidy profit to the innovator. However, all too often the marketing mechanism employed is not *marketing pull*, but *technological push* (Herbig and Day, 1992). Marketers often presume that since the technology exists and an innovation has been created, its diffusion is inevitable—a *fait accompli*. As we have seen in the past, successful innovations are often those which pay more attention to market demand than to technological opportunity.

A product or service must be relevant, have demonstrated value and meet specific needs in order to prosper. Therefore, the ultimate objective of any commercial research and development program should be to secure a better match between the production of resources and their utilization by industry and consumers. As eluded to earlier, public reaction will be vital in determining overall market impacts of animal biotechnology. Consequently, greater effort must be focused toward cost–reducing and environmentally friendly innovations. While there are many promising applications of biotechnology on the horizon, biotechnology is neither a panacea nor a complete replacement for established tools. It provides an additional approach to agricultural problems.

Recent studies have shown, among other things, that emerging products of biotechnology will require considerable management expertise on the part of producers (National Research Council, 1990; Office of Technology Assessment, 1991d; Van der Wal et al., 1991). As with many other technological advances, the farmers that will benefit most will be the more efficient managers and early adopters. Furthermore, price support programs, marketing orders, grading systems and regulatory mechanisms will all need to adapt to tomorrow's dynamic production systems.

LOOKING TO THE FUTURE

Leadership in technology development and utilization is the role the U.S. has, can, and seeks to assert for the rest of the world. As noted earlier, the U.S. federal investment in biological research of the past 30 years has laid the foundation for a strong biotechnology enterprise. Because the field is moving rapidly, historical leadership does *not* ensure continued superiority (Federal Coordinating Council for Science, Engineering, and Technology, 1992).

Looking toward the future, what elements could present a positive sum strategy for animal biotechnology in the 1990s? What will be the challenges that influence animal agriculture in the years to come?

First of all, a sound national strategy for biotechnology in agriculture must focus on solving pivotal scientific and agricultural problems, effectively utilizing the funds and institutional structures available to support research, training researchers in advanced scientific areas, and efficiently transferring technology (National Research Council, 1987). Both industry and government have appropriate roles to play in this process (Council on Competitiveness, 1991). There is a need to construct institutional infrastructures that facilitate more effective collaboration among animal scientists, engineers, agronomists and health scientists to deal with issues of production, environmental change and the health of producers and consumers (Ruttan, 1991).

The Cooperative Extension Service and educational institutions must keep pace with ongoing change to be relevant to the future competitiveness and profitability of American agriculture. Producers of the future will need, and desire, a new menu of technologies that recognize contemporary goals such as enhanced profitability, increased environmental stewardship, rural revitalization and development, and global competitiveness (National Research Council, 1987). Further, there is a need to increase the involvement of farmers, researchers and allied industry in developmental partnerships. These challenges will demand a correspondingly higher level of vision and sophistication on the part of government policymakers, regulators and industry leaders.

Crucial to all the various points discussed previously is the effectiveness of our world trading system. A global approach to the regulation and acceptability aspects of biotechnology is worth pursuing in order to create an improved atmosphere of mutual confidence between producers, manufacturers and consumers. Moreover, the removal of nontariff trade barriers between the world's trading partners and the development of a common reference point is of vital importance while at the same time providing the flexibility to accommodate unforseen and justifiably unique national considerations.

CONCLUSIONS

World agriculture stands at the threshold of new scientific and technical developments in animal science, biology, chemistry, genetics, agricultural engineering, information technology and many other fields. In most of the world, the transition from a resource–based to a science–based system of agriculture is occurring within a single century (Ruttan, 1991).

Emerging technologies, industry economics and public policy will play critical roles in shaping U.S. animal agriculture in the decade of the 1990s. Advances in health maintenance, reproduction efficiency and information technology will all affect the industry. Additional research is needed to gain an increased understanding of the factors influencing animal growth, environmental adaptation and well–being, and disease resistance (National Research Council, 1989).

Legislative and regulatory activities that occur in Washington are having a greater effect on animal agriculture each year. A simple statement of need is no longer enough to justify the allocation of funds for new programs, new facilities or new research efforts (Waggoner et al., 1989). Consultancies, affiliate programs, consortia, research parks and other forms of partnership between the public and private sectors that foster communication and technology transfer should be promoted (National Research Council, 1987).

Certainly, we must not overlook or push aside the legitimate concerns of the public and work to establish those principles which govern the safe environmental use of emerging products. In such a fast–moving technological environment, it will be necessary to regularly review the appropriateness of the scientific basis of existing regulation and to make any required adjustments in either the technology or the statutory framework.

There are, of course, numerous scientific and technological bottlenecks and data gaps that still have to be overcome, as can be expected of a technology that has been expanding so swiftly and in so many directions. Granted, the outcome of the best science can be unpredictable. The obstacles to crafting an effective strategy to support competitiveness in animal biotechnology are formidable. However, the potential payoffs are abounding.

The genius that has driven America's prosperity throughout its history has been the ability to combine collective vision with diversity and individualism—to unite grand ideals with arduous pragmatism. As U.S. agriculture enters the 21st century, this genius will be put to its most strident test.

REFERENCES

Burrill, G.S. and K.B. Lee, Jr. 1991. *Biotechnology '92: Promise to reality.* Ernest & Young, San Francisco, CA.

Council on Competitiveness, Office of the Vice President. 1991. *Report on national biotechnology policy.* U.S. Government Printing Office, Washington, DC.

Federal Coordinating Council for Science, Engineering, and Technology. 1992. *Biotechnology for the 21st century: The FY 1993 U.S. biotechnology research initiative.* A Report from the Committee on Life Sciences and Health. U.S. Government Printing Office, Washington, DC.

General Accounting Office. 1992. *Food safety and quality: Uniform, risk–based inspection system needed to ensure safe food supply.* GAO/RCED–92–152. U.S. Government Printing Office, Washington, DC.

Harlander, S.K. 1991. Communication between scientists and consumers. *Outlook on Agriculture.* 20(2):73–77.

Herbig, P.A. and R.L. Day. 1992. Customer acceptance: The key to successful introductions of innovations. *Marketing Intelligence & Planning.* 10(1):4–15.

Kalter, R.J. 1985. The new biotechnology agriculture: Unforeseen economic consequences. *Issues in Science and Technology.* 2:125–133.

Kopchick, J.J. 1992. Biotechnology and food safety. *Journal of the American Veterinary Medical Association.* 201(2):228–234.

Moses, V. and R.E. Cape, eds. 1991. *Biotechnology: the science and the business.* Harwood Academic Publishers, London.

National Research Council. 1987. *Agricultural biotechnology–strategies for national competitiveness.* National Academy Press, Washington, DC.

National Research Council. 1989. *Investing in research: A proposal to strengthen the agricultural, food, and environmental system.* National Academy Press, Washington, DC.

National Research Council. 1990. *Technology and agricultural policy: Proceedings of a symposium.* National Academy Press, Washington, DC.

Office of Technology Assessment. 1988. *Technology and the American economic transition: choices for the future.* OTA–TET–283. U.S. Government Printing Office, Washington, DC.

Office of Technology Assessment. 1989. *New developments in biotechnology: Patenting life–special report.* OTA–BA–370. U.S. Government Printing Office, Washington, DC.

Office of Technology Assessment. 1991a. *Biotechnology in a global economy.* OTA–BA–494. U.S. Government Printing Office, Washington, DC.

Office of Technology Assessment. 1991b. *Competing economies: America, Europe, and the Pacific Rim.* OTA–ITE–498. U.S. Government Printing Office, Washington, DC.

Office of Technology Assessment. 1991c. *Federally funded research: Decisions for a decade.* OTA–SET–490. U.S. Government Printing Office, Washington, DC.

Office of Technology Assessment. 1991d. *U.S. Dairy industry at a crossroad: Biotechnology and policy choices.* OTA–F–470. U.S. Government Printing Office, Washington, DC.

Pearson, A.M. and T.R. Dutson, eds. 1990. *Meat and health: Advances in meat research, Vol. 6.* Elsevier Applied Science, London.

Raines, L.J. 1991. Protecting Biotechnology's Pioneers. *Issues in science and technology.* 8(2):33–39.

Ruttan, V.W. 1991. Constraints on sustainable growth in agricultural production: Into the 21st century. *Outlook on Agriculture.* 20(4):225–233.

Stenholm, C.W. and D.B. Waggoner. 1992. Moving beyond the rhetoric of food safety and meeting the challenge. *Journal of the American Veterinary Medical Association.* 201(2):234–239.

Van der Wal, P., G.M. Weber and F.J. Van der Wilt, eds. 1991. *Biotechnology for control of growth and product quality in meat production: Implications and acceptability.* Proc. of an International Symposium. Pudoc, Wageningen.

von Oehsen, W.H, III. 1988. Regulating genetic engineering in an era of increased judicial deference: A proper balance of the federal powers. *Administrative Law Review.* 40(3):303–344.

Waggoner, D.B., D.K. Waggoner, J.R. Lyons and T.L. Nipp. 1989. The importance of effective Washington representation for animal and dairy science: Developing the proper presence. *Journal of Animal Science.* 67(Suppl.1):616.(Abstr.)

Wald, S. 1992. Biotechnology, agriculture and food. *The OECD Observer.* 177:4–8.

Yuan, R.T. and M.D. Dibner. 1990. Japanese biotechnology: A comprehensive study of government policy, R&D and industry. Stockton Press, New York.

Neal L. First
Animal Science
University of Wisconsin

Animal Biotechnologies: Potential Impact on Animal Products and Their Production

Biotechnologies being developed for use in animal agriculture include the commonly practiced technologies of artificial insemination and embryo transfer, as well as the developing technologies associated with *in vitro* production of embryos, the splitting and cloning of embryos, marker–assisted selection including sexing of the embryo, and the transfer of new genes into an embryo. Each technology should be considered separately when assessing the benefits and risks of each to animals and humans because each is distinctly different and only one, gene transfer, involves recombinant DNA technology.

Genetic improvement of farm animals by traditional parent selection has been slow, especially for traits of low heritability. Nevertheless, the rate of increase in milk production has been greatly accelerated by artificial insemination of cows with semen from highly–selected, performance–tested bulls. In a limited way, the valuable genetics of a few very high production cows has been extended several–fold by the use of superovulation and embryo transfer. Conventional mating, artificial insemination and embryo transfer all have the disadvantage of propagating undesirable genes of a high–performance male or female along with the selected genes. Exciting new developments in animal biotechnology offer hope for modeling and designing animals to fit market and environmental needs and for rapidly propagating or identifying the animals of superior performance. This review will focus on each of these biotechnologies, their development status, use or potential use in animal agriculture, benefits and risks to the consumer of animal products and risks to animals.

ARTIFICIAL INSEMINATION

Use of artificial insemination developed rapidly in dairy cattle beginning in the late 1940s and 1950s when it was realized that bull sperm could be stored frozen and that 2–4 ejaculates per week from a bull could provide sufficient sperm after frozen storage to inseminate at least 2,000 cows. This allowed extreme selection of the sires, resulting in a more than two–fold increase in genetic ability for milk production of dairy cows. At present the pregnancy rates are 60–65 percent and most dairy cows are mated by artificial insemination (70 percent U.S., greater than 90 percent Europe). While this technology has been developed for use in other domestic animals, the cost–benefit ratio has been favorable in the U.S. only for wide–scale use in dairy cattle or use by breeders at the top of the breeding stock pyramid for beef cattle, sheep and swine. It is a common practice in poultry breeding. The benefit to consumers has been low cost and availability of dairy products of high quality. The risks to the animals are essentially none. While it is sometimes argued that high milk production could reduce cow longevity, older cows declining in milk production are usually slaughtered for meat before becoming aged. For review of artificial insemination see Hafez, 1987.

EMBRYO TRANSFER

Embryo transfer (Figure 1) is being used primarily in dairy cattle and the top seedstock herds of beef cattle. It has the advantage of genetic improvement through both sire and dam rather than the sire alone, as with artificial insemination. There are approximately 250,000 calves born annually in the U.S. from embryo transfer. Its use has been limited because the technology of superovulation and embryo transfer has allowed only 20–30 calves per year from a cow and because it is more expensive than artificial insemination.

FIGURE 1

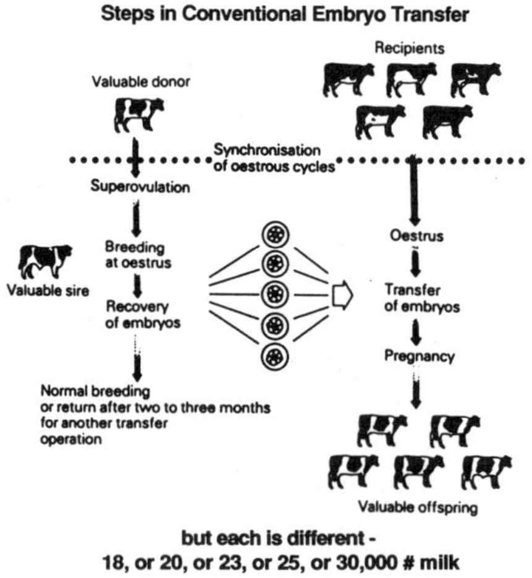

Steps in Conventional Embryo Transfer

Recipients
Valuable donor
Synchronisation of oestrous cycles
Superovulation
Breeding at oestrus
Valuable sire
Recovery of embryos
Oestrus
Transfer of embryos
Pregnancy
Normal breeding or return after two to three months for another transfer operation
Valuable offspring

but each is different -
18, or 20, or 23, or 25, or 30,000 # milk

Blastocyst stage embryos are recovered nonsurgically by flushing the uterus of a superovulated, genetically superior cow that had been inseminated with sperm from a bull of high genetic value. The embryos numbering 3 to 10 or more are transferred each to a recipient cow synchronized in estrous cycle with the donor. The embryos can be stored frozen or sexed. Each calf born is no more alike than usual siblings and for traits such as milk production, where heritability of the trait is low, only a few reach the level of production of the parents.

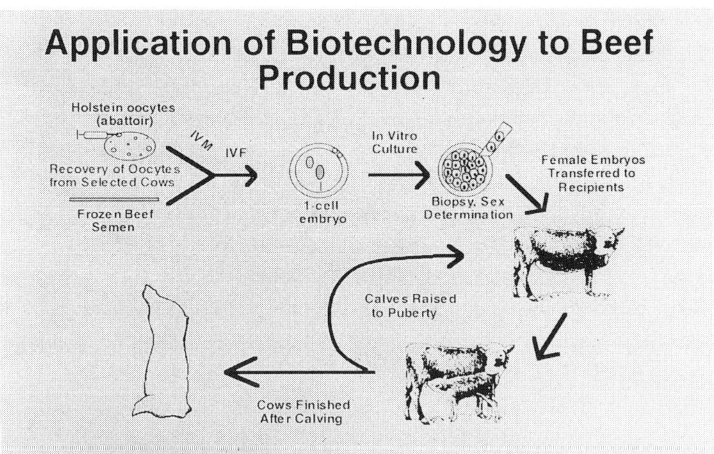

FIGURE 2

One commercial use for 'in vitro' production of embryos is illustrated above (Schaefer et al., unpublished). This breeding plan is used to eliminate maintenance of a beef brood cow. Other uses when oocytes are recovered by transvaginal laparoscopy from valuable cows include the production of large numbers of offspring from a valuable cow, or old or incapacitated cows.

Embryos can, and are, being frozen, and in a few cases sexed or split to double the number of embryos. Procedures for superovulation and embryo transfer are nonsurgical and present little risk to the cow donating the embryo or the recipient receiving it and cows can be successfully superovulated or bred at the second estrus cycle after superovulation. The animal risk is low and the benefit to consumers is manifest as more abundant, lower cost, higher quality milk and meat products. For review of embryo transfer see First, 1991; Seidel, 1991; and Wilmut et al., 1992. Artificial insemination and especially embryo transfer are the delivery mechanisms by which the new reproductive and genetic biotechnologies now under development will be delivered to animals for propagation.

IN VITRO PRODUCTION OF EMBRYOS

In vitro production of embryos depends on efficient systems for culture of oocytes and sperm, fertilization and embryo culture. This technology is being developed for all food–producing animals and is presently best developed for cattle where it is beginning to be applied. Parts of this technology are essential for cloning of embryos and for gene transfer. Several breeding companies are applying this technology commercially. Its application has been in two forms. In field trials in the U.S., Japan and Great Britain, embryos are produced *in vitro* from abattoir–recovered oocytes of selected breeds that are fertilized with semen of highly selected bulls. This is done with the objective to replace the brood cow in beef production with *in vitro* produced embryos (Schaefer et al., unpublished) as shown in Figure 2. Each group has used this approach for a different genetic purpose. The application in Japan is to use, for both reproduction and valuable beef, the young Wagyu females that produce expensive Kobi beef. In Britain, the use is to produce valuable beef from dairy cows; while the use in the U.S. is to produce both embryo transfer recipient females and beef from young females derived from *in vitro* production of embryos, as shown in Figure 2 (Schaefer et al., unpublished). Cattle oocytes can also be recovered from follicles matured *in vivo* by recovery using ultrasound–guided vaginal laparoscopy. Recovery of these oocytes from genetically valuable cows provides a supply of oocytes of high genetic value for use in gene transfer and production of oocytes and embryos for cloning. Recent estimates indicate that one genetically valuable cow could produce 10 oocytes every 2–3 days throughout the year (Krimpenfort et al., 1991; Van der Schans et al., 1991) or approximately 100 calves per year, a big increase from that achieved by superovulation. Several embryo transfer companies are preparing to offer this *in vitro* embryo production service. The major challenge to researchers is to harvest and mature the thousands of growing oocytes and small follicles of domestic species. This would further increase the pool of oocytes available from genetically valuable animals. Application of this technique to fetal ovaries would allow rapid genetic progress through marker–assisted selection and velogenesis (Georges, 1991). The second part of producing embryos *in vitro* is the sperm capacitation and fertilization system. In general, any agent that causes Ca^{++} entry into the sperm acrosome and a pH increase within the sperm causes capacitation. Numerous capacitation systems have been developed including high ionic strength media, glycosaminoglycans (such as heparin), aging, pH shift, calcium ionophores, caffeine and oviduct fluid. With appropriate sperm capacitation and incubation in serum–free medium at body temperature, *in vitro* fertilization rates have been reported as high as 70–80 percent in cattle, sheep, swine and goats (Parrish, 1990; First, 1991). Embryos of domestic animals can be successfully cultured in surrogate oocytes of rabbits and sheep or cultured with oviduct cells or oviduct cell conditioned media and recently successfully cultured in a defined media (Rosenkrans and First, 1990; First, 1991). While investigators are still searching for uterine factors and growth factors that may further increase the survival and development of cultured embryos, present methods are satisfactory for commercial use. The *in vitro* production of embryos as now practiced presents several benefits to society other than a lower–cost, more–abundant food product. This technology is economical to the environment and food supply as it eliminates beef brood cows. It permits use of gametes, cells and embryos in research rather than use of animals. To the livestock producer it could allow cost–effective genetic improvement. The risks are minimal when oocytes are recovered transvaginally and none when they are recovered as an abattoir by–product. Pregnancy rates are approximately

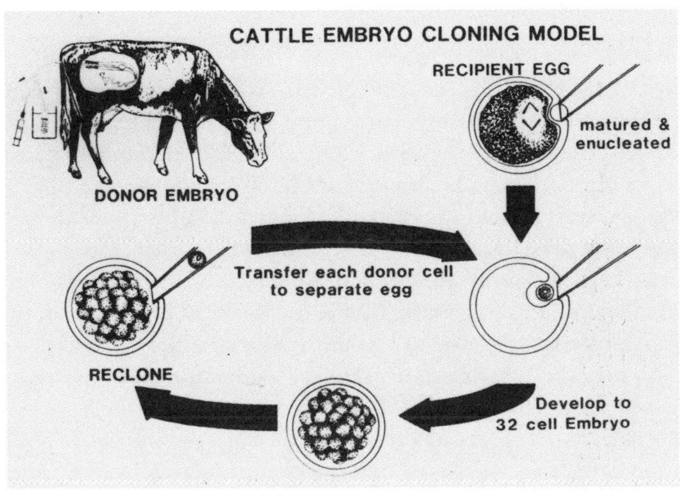

FIGURE 3

Donor embryos are obtained by flushing the uterus of a genetically superior cow after superovulation and insemination with sperm from a genetically superior bull. Each cell of each 30 to 60 cell stage embryo is transferred into an enucleated oocyte. The embryos produced by this process can be used to supply donor cells to further increase the number of cloned embryos by recloning.

---

60 percent and offspring are normal (Monson et al., 1992). For a review of this technology, see Leibfried–Rutledge el al., 1989; Gordon and Lu, 1990; First, 1991; Hansel and Godke, 1992. These same methods for *in vitro* production of embryos are also beginning to be used to propagate zoo animals and to save endangered species (Wildt et al., 1992).

CLONING OF DOMESTIC ANIMALS

Twins are presently being produced with good efficiency by bisection of embryos and a few cells left on the bisection knife have been used to sex the bisected embryos. The most promising method for production of large numbers of offspring is nuclear transplantation. This procedure has successfully produced viable embryos and offspring in cattle, sheep, rabbits and swine. The procedure (Figure 3) involves transfer of a blastomere or nucleus from the valuable embryo at a multicellular stage (usually 20–120 or more cells) into an enucleated metaphase II oocyte. The oocyte then develops to a multicellular stage and is used as a donor in a serial recloning (First and Prather, 1991; Stice, 1992). Nuclear transplantation is being developed in private industry as well as by university research. Thus far, nuclear transplantation in cattle has been successfully performed using low–cost, *in vitro* matured oocytes from abattoir–recovered ovaries and with serial nuclear transfers. However, the efficiency is less than desired with approximately 20–25 percent of the nuclear transplantations resulting in transferable embryos and approximately

30 percent of the embryos transferred into the cows resulting in completed pregnancies. Throughout the U.S. and Canada, several hundred pregnancies have been produced in cattle by this procedure and recloning has been performed. To date, the largest number of calves cloned from one embryo has been 11 calves born at Granada Genetics in 1990 (First and Prather, 1991). The keys to a successful cloning system for a livestock industry are the ability to use donor embryos of larger cell numbers to produce many offspring and the capacity to use cells from cloned embryos as the donor nuclei for another generation of clones. In sheep embryos, the frequency of development to blastocysts after use of donor cells from the blastocyst inner cell mass was 57 percent and pregnancies resulted. In rabbits, blastocysts have been produced from inner cell mass cells but at a lower frequency than from the 32–cell morula stage blastomeres. In cattle, embryos at the stages of morula or the inner cell mass of blastocysts have produced good results as donors in cloning. This is approximately the stage where embryonic stem cells can be recovered and multiplied in culture in a mouse. If similar stem cell isolation and multiplication were done in domestic animals and if stem cells should prove useful in cloning by nuclear transplantation, the number of possible clones is unlimited (First, 1991). When developed to high efficiency, cloning provides a nearly phenotypic selection and propagation system for replicating valuable animals. For example, traits with heritabilities of approximately 30 percent are expected to increase to nearly 70 percent. It will also be used for rapid propagation of precious transgenic animals. The benefits of nuclear transfer include a nearly phenotypic selection, accelerated genetic improvement or environmental adaptation and characterized and predictable production performance, nutrient requirements, disease resistance and extensive screening of clonal lines for genetic defects, disease resistance and environmental adaptation before multiplication and release for use. The risks are low to none for the animals supplying donor embryo cells or the recipient oocytes, but the process at the present state of the art results in less than normal embryo survival before embryo transfer and less than normal pregnancy rates and calving rates. Also, some of the calves are born larger than normal and require assisted delivery. It is expected that with time and continued research these problems will be understood and corrected. For reviews of cloning of domestic animals, see First and Prather, 1991; Bondioli, 1992; Prather et al., 1992; and Seidel, 1992.

GENE TRANSFER

Successful production of transgenic food–producing animals requires the ability to efficiently achieve development of the transgenic embryo and genome integration of the transgene. Precise genetic modeling and appropriate promoter sequences to achieve expression at a high level in a tissue of choice and of the trait desired are necessary. Transgenic cattle, sheep, swine and rabbits have been made by microinjection of DNA into a pronucleus of a one–

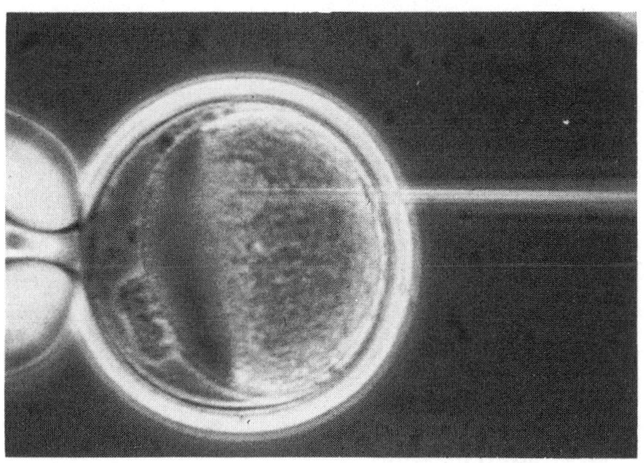

FIGURE 4

Microinjection of concentrated DNA (a gene) into a pronucleus of a bovine egg. Note the egg is first centrifuged to concentrate the cytoplasm against one side such that the pronuclei can be visualized in the cleared area.

cell zygote (Figure 4), transgenic fish by injection of DNA into oocytes and chickens by infection of genes into eggs (Hansel and Weir, 1990; First and Haseltine, 1991; Rexroad, 1992). Principle genes introduced into fish have been growth hormone genes resulting in production of fish that grow 2–3 times faster than normal. Genes imparting cold resistance to warm water species have also been expressed by transgenic fish. Transgenic chickens have been made which express increased growth from a growth hormone transgene and increased viral resistance from interference with cell receptors for avian viruses. Cattle, sheep and swine have been made transgenic for various growth hormones without significant increases in growth, but with decreased fat in the carcass. Use of tissue–specific promoter sequences and appropriate promoter control of the level of gene expression should improve growth responses. The SKI oncogene is an example of a gene that enhances muscle growth in mice and swine. Other genes may be identified that increase muscle tenderness. When appropriate disease resistance genes are identified, it should be possible to engineer high–producing animals for survival in high–disease environments. Genes for expanding the MHC complex and globins have been introduced into mice and sheep. Cells of animals have been genetically modified to resist the entire herpes family and defective viruses have been used in chickens to promote receptor resistance to pathogenic viruses (Hansel and Weir, 1990; First and Haseltine, 1991; Rexroad, 1992). The ability to target gene expression exclusively to the mammary gland will allow modification of milk composition to make novel cheeses, remove milk fat, lactose or allergenic proteins and increase protein content. Thus far, transgenic mice have been made which express new caseins or no milk fat. It is likely that some cows will be designed to produce milk for specialty dairy products while most cows may be engineered to produce little or no fat in their milk. It is also expected that pharmaceutical products will be produced from milk of cows expressing pharmaceutical transgenes in their mammary glands. Already transgenic mice, sheep, goats and pigs have been produced which expressed either the pharmaceutical proteins of tissue plasminogen activator clotting factor 9, alpha–l–antitrypsin, lactoferrin, eurokynase, follicle–stimulating hormone, protein C, human growth hormone or interleukin 2 in their milk (First et al., 1991; Rexroad, 1992). Gene transfer usually results in one or two transgenic animals forming the beginning of a transgenic line. It, therefore, does not initially impact a large part of a population and requires artificial insemination, or *in vitro* production of embryos, or cloning or combinations of the above to produce animals or fish which are commercially useful. Gene transfer is most useful for introduction of genes not found in a population or deletion of genes not wanted, whereas marker–assisted selection is considered to be much more efficient in changing a population if the gene in question exists in the population. The benefits from gene transfer other than increasing the efficiency of animal production include the development of animals better fit for specific environments, including disease resistance; the production of new animal products and higher quality products, for example more digestible milk; removal of allergenic compounds from milk; etc. The technique of gene transfer imposes no direct animal risk. However, a greater than normal early loss of embryos occurs. The cost of gene transfer and use of animals can be considerably reduced by transfer of genes into *in vitro* produced embryos (First et al., 1991; Krimpenfort et al., 1991). Faulty modeling of the gene and promoter construct can result in insertion of the transgene at an inappropriate site with disturbance of expression of another gene or expression in other than the tissue or cell targeted in the modeling of the transgene, or expression at an inappropriate time in animal development. In the future, these risks will likely be reduced or eliminated by advances in several areas. These include the introduction of DNA into cultured cells that can be sampled and screened for appropriate expression before use in nuclear transfer to make embryos for transfer into cows. The use of cultured cells to make animals also allows site specific gene transfer or deletion through homologous recombination, thereby eliminating inappropriate sites of integration. Improved modeling of the desired outcome of the transfer gene will occur as gene mapping projects provide genome knowledge sufficient to allow accurate modeling of the genome and the gene construct. Perhaps the greatest challenge will be the development of consumer confidence that specific genetically engineered animals may not be at risk and that engineered products are safe whereas other transgenics may be rejected because animals are at risk. For review of gene transfer see Rexroad, 1992; First et al., 1991; First and Haseltine, 1991; Wilmut et al., 1990; and Hansel and Godke, 1992.

MARKER–ASSISTED SELECTION

Efforts to map the genomes of domestic animals and similarities with the mapped human genome have resulted in DNA markers that are beginning to be associated with desired or undesired productivity traits. For example, a restriction fragment linked polymorphism (RFLP) in at least one family of Holstein cattle has been associated with high milk production. Several artificial breeding companies now use this DNA marker to select for higher milk production. Markers for K–casein that relates to protein content of the milk and markers for selection against a neurological defect in Brown Swiss cattle are also in use (Georges, 1991). Rapid development of markers for use in phenotypic and genetic selection is expected as more of the genome and specific linkages to production traits are understood. Because the genomes of higher mammals are similar and gene mapping efforts across species are coordinated, we are rapidly increasing our knowledge of the genome. The applied value of gene mapping and genetic markers is primarily through association of markers with productivity, disease resistance and product quality: traits of interest. But marker–assisted selection has several other advantages. Markers can be used to perform early genetic selection on gametes, embryos or newborn animals. Markers can be used for DNA fingerprinting and accurate animal identification or association of product with animal, herd or processing plant. Markers are used to screen for genetic defects and when genes are introduced from other populations, markers can be used to track their segregation in the population. Because marker–assisted selection imposes little to no risk to the animals donating blood, sperm, or embryos for assay and imposes no risk to the consumer, marker–assisted selection is expected in the short term to be the most commonly used of the above animal biotechnologies. For review of marker–assisted selection see Dentine, 1992; Georges, 1991; Fries et al., 1989; and Massey and Georges, 1992. Several other biotechnologies impact animal agriculture, but also are expected to impose no risk to animal or consumer. These include the use of DNA fingerprinting for diagnosis of disease microorganisms, the development of new vaccines using recombinant technology and the use of engineered colostrums to protect animals and humans from disease.

SUMMARY

In summary, it is apparent that every biotechnology is different. Each gene construct is different and each must be examined for its individual benefits and risks. Some biotechnologies reduce the need for animals in research or reduce the numbers needed for food production. Some protect the health of the animal or make it more fit for a changed environment and some allow for the preservation and rapid repopulation of an endangered species. Most importantly, biotechnology is a series of tools to be used intelligently or carelessly by humans as a choice of humans. We must choose wisely, but avoid condemnation and rejection of the tool. Like fire, clothing and the wheel, we may someday need it and be wise enough to use it.

REFERENCES

Bondioli, K.R. 1992. *Commercial cloning of cattle by nuclear transfer.* Proc. Symposium on Cloning Mammals by Nuclear Transplantation (January 15th) Fort Collins, CO. pp. 35–38.

Dentine, M.R. 1992. Marker–assisted selection in cattle. *Anim. Biotech.* 3(1):81–93.

First, N.L. 1991. New advances in reproductive biology of gametes and embryos. In *Animal Applications of Research in Mammalian Development.* R.A. Pedersen, A. McLaren, and N. First, eds. Cold Springs Harbor Laboratory, New York. pp. 1–21.

First, N.L. and F.P. Haseltine. 1991. *Transgenic Animals.* Butterworth–Heinemann, Boston.

First, N.L. and R.S. Prather. 1991. Genomic potential in mammals. *Differentiation.* 48:1–8.

First, N., T. Kim, G. McNamara and G. Bleck. 1991. *Transgenic animals.* The 2nd ARRC International Symposium, Korea. pp. 9–32.

Fries, R., J.S. Beckmann, M. Georges, M. Soller and J. Womack. 1989. The bovine gene map. *Animal Genetics.* 20:3–29.

Georges, M. 1991. Perspectives for marker–assisted selection and velogenetics in animal breeding. In *Animal Applications of Research in Mammalian Development.* R.A. Pedersen, A. McLaren, and N. First, eds. Cold Springs Harbor Laboratory, New York. pp. 285–325.

Gordon, I. and K.H. Lu. 1990. Production of embryos 'in vitro' and its impact on livestock production. *Theriogenology.* 33:77–88.

Hafez, E.S.E. 1987. Artificial insemination. In *Reproduction in Farm Animals.* E.S.E. Hafez, ed. Lea & Febiger, Philadelphia. pp. 481–497.

Hansel, W. and B.J. Weir, eds. 1990. *Genetic Engineering of Animals.* Proceedings of the Second Symposium on Genetic Engineering of Animals held at Cornell University, Ithaca, NY.

Hansel, W. and R.A. Godke. 1992. Future prospectives on animal biotechnology. *Anim. Biotech.* 3 (1) 111–137.

Krimpenfort, P., A. Rademakers, W. Eyestone, A. Van der Schans, S. Van den Broek, P. Kooiman, E. Kootwijk, G. Plantenburg, F. Pieper, R. Strijker and H.A. de Boer. 1991. Generation of transgenic dairy cattle using 'in vitro' embryo production. *Bio/Technology.* 9:844–847.

Leibfried–Rutledge, M.L., E.S. Critser, J.J. Parrish and N.L. First. 1989. 'In vitro' maturation and fertilization of bovine oocytes. *Theriogenology.* 31:61–74.

Massey, J.M. and M. Georges. 1992. Genmark's approach to marker–assisted selection. *Anim. Biotech.* 3(1):95–109.

Monson, R., D.L. Northey, R. Gottfredson, D.R. Peschel, J.J. Rutledge and D.M. Schaefer. 1992. Pregnancy rates of 'in vitro' produced bovine embryos following nonsurgical transfer. *Therigenology.* 37:261 (abstract).

Parrish, J.J. 1990. Application of 'in vitro' fertilization to domestic animals. In *The Biology and Chemistry of Mammalian Fertilization, Vol. II.* CRC Press, New York. pp. 111–132.

Prather, R.S., T.T. Stumpf and L.F. Rickords. 1992. Nuclear Transplantation as a method of producing genetically identical livestock. *Anim. Biotech.* 3(1):67–79.

Rexroad, C.E., Jr. 1992. Transgenic technology in animal agriculture. *Anim. Biotech.* 3(1):113.

Rosenkrans, C.F., Jr., Zeng, G.Z., Schoff, P.K. and N.L. First. 1990. A simple medium for 'in vitro' development of bovine embryos. *J. Anim. Sci.* 68(1):430 (abstract)

Rosenkrans, C.F. and N.L. First. 1991. Culture of bovine zygotes to the blastocyst stage: Effects of amino acids and vitamins. *Theriogenology.* 35:266 (abstract).

Seidel, G.E. Jr. 1991. Embryo transfer: The next 100 years. *Theriogenology.* 35:171–180.

Seidel, G.E. Jr. 1992. *Overview of cloning mammals by nuclear transplantation.* Proc. Symposium on Cloning Mammals by Nuclear Transplantation January 15th. Fort Collins, CO. pp. 1–4.

Schaefer, D., J.J. Rutledge; D. Northey; R. Munson and D. Peschal. unpublished.

Stice, S.L. 1992. *Multiple generation bovine embryo cloning.* Proc. symposium on cloning mammals by nuclear transplantation. January 15th. Fort Collins, CO. pp. 28–31.

Van der Schans, A., L.A.J. Van der Westerlaken, A.A.C. de Wit, W.H. Eyestone and H.A. deBoer. 1991. Ultrasound–guided transvaginal collection of oocytes in the cow. *Theriogenology.* 35(1):288 (abstract).

Wilmut, I., A.L. Archibald, S. Harris, M. McClenaghan, J.P. Simons, C.B.A. Whitlaw and A.J. Clark. 1990. Methods of gene transfer and their potential use to modify milk composition. *Theriogenology.* 33:113.

Wilmut, I., K.H.S. Campbell and G.T. O'Neill. 1992. *Sources of totipotent nuclei including embryonic stem cells.* Proc. Symposium on cloning mammals by nuclear transplantation. January 15th. Fort Collins, CO. pp. 8–16.

Wildt, D.E., S.L. Monfort, A.M. Donoghue, L.A. Johnston and J. Howard. 1992. Embryogenesis in conservation biology–or, how to make an endangered species embryo. *Theriogenology.* 37(1):161-184.

Michael W. Fox
Vice President, Bioethics and Farm Animals
The Humane Society of the United States

The New Creation: An Update on Animal Gene Engineering

There have been several new developments in genetic engineering that show how this new industry is applying biotechnology in agriculture and medicine. How valuable these new developments are—in terms of real progress in improving agricultural practices and human health—remains to be seen. The following examples clearly reveal that the "New Creation" and new world order of the biotechnology industry is far from any utopian dream of a world made perfect for humankind. One can read between the lines of the new patent applications, news releases and scientific reports concerning the latest feats of genetic engineering and glimpse into the future. The wonder–world of the New Creation is not quite here today, but it may be upon us sooner than we expect. A whole new generation of genetically engineered (so–called transgenic) animals is on the horizon. These will carry genes taken from humans and other species. In the world of trade and commerce, they will be regarded as "new" species—unique, patentable commodities of the new world order.

TRANSGENIC ANIMALS

Scientists in the U.S., Japan, Europe and Australia have created a number of transgenic animals—pigs, lambs, calves and fish—containing the genes of other species like the human and bovine growth hormone genes. Success rates of gene insertion are extremely low and the entire process is time consuming and costly. Much of the funding in this area of biotechnology comes from the public via government tax revenues.

Some researchers have recently opted to put extra growth–regulating genes of sheep origin into lambs rather than human genes because they felt that "transgenes composed entirely of sheep gene sequences would be more acceptable to lay persons, in particular consumers" (Murray and Rexroad, 1991). However, even though these lambs were leaner, they did not have increased feed efficiency. They were diabetic and had such severe health problems that they died before reaching puberty. "The cause of death has varied, but there are clear data that the over–expression of GH (growth hormone) adversely affects liver, kidney and cardiac function" (Murray and Rexroad, 1991).

Merck & Co., the European–based pharmaceutical company, has applied for a patent in Europe on its superchicken, or Macro–Chicken (Holden, 1991). They have developed a line of broiler chickens that carry the growth gene of cows with the hopes of cornering the market with a highly feed–efficient, fast–growing bird.

It is likely that Merck's Macro–Chickens will have a variety of health problems too. But if the birds eat well and grow quickly, they will be ready for slaughter before severe health problems ever develop. But what of the breeding stock of transgenic chickens that will not be raised for slaughter? Will they suffer? Because such information is proprietary, corporations are not likely to reveal the limitations and risks of their new patentable creations. Trade secrets notwithstanding, the social and economic consequences—to farmers, to the practice and structure of agriculture and to consumers—of creating transgenic farm animals have been given scant attention.

Critics of the genetic engineering of farm animals question the use of public funds to make these animals produce more meat (even if it *is* lean) when the short– and long–term costs of such research are not considered and when a major problem of contemporary intensive animal agriculture is overproduction, meat and milk surpluses being a chronic problem. It is unlikely that the creation of transgenic farm animals will help feed the hungry world since meat production efficiency has built–in limitations and inevitable environmental costs (Durning and Brough, 1991; Fox, 1990).

Genetic engineering technology is being used in an attempt to alter sheep's and cows' milk so that it can be consumed by a large percent of the world population that is lactose–intolerant (Mercier, 1987). This may be a more fruitful approach to helping feed the hungry, since milk production is far more efficient, ecologically sound and cost–effective than meat production, with or without biotechnology.

Human genes responsible for the production of proteins in mothers' milk are being inserted into calf embryos with the hope of creating a new generation of cows that produce "humanized" milk (Phelps, 1989).

Australian government scientists are using genetic engineering to make sheep produce more wool by inserting genes into their developing embryos. The sheep's body chemistry is altered to convert sulfur–bearing compounds into methionine, an amino acid that increases wool growth (Ford, 1988). Australian scientists are also trying to genetically engineer a hormone that can be injected into sheep that will make them shed their fleece, thus cutting down costs of shearing. Tests to date have caused pregnant sheep to abort (Scherer, 1992; *New Scientist*, 1992). They also plan to genetically engineer sheep that secrete insect repellent from their hair follicles to ward off blowflies. Blowflies cost the sheep industry $85 million per year in losses. As a spinoff, the sheep will also have the world's first moth–proof wool.

It should be emphasized that most genetic engineering research in farm animals has focused on increasing productivity, while research on increasing resistance to disease through genetic engineering (Slater, 1990; Kraemer and Templeton, 1990) is still very much in its infancy. This latter area of research should be questioned since improvements in farm animal husbandry are surely more cost–effective ways of improving animal health and well–being.

TRANSGENIC "MOLECULAR PHARMING"

Human genes are being inserted into farm animals so that they produce various pharmaceutical products in their milk, such as blood clotting factors and other substances of possible medical application (Clark et al., 1987; Watts, 1990; Schanbacher, 1990; Bialy, 1991). Harvey Bialy, editor of *Bio/Technology*, has extolled the virtues of what he terms "molecular pharming technologies," as exemplified by research teams from the UK, U.S. and The Netherlands, who have produced transgenic sheep whose milk contains human alpha–1–antitrypsin, transgenic goats that secrete tPA into their milk and the first transgenic dairy cattle. "Taken together," he writes, "their results provide a convincing demonstration of the feasibility of using animals as commercial bioreactors" (Bialy, 1991). It will be many years before these new animal creations provide any medical benefits to humans, but venture capitalists are investing in this speculative line of research and development. Recently, a biotechnology company, DNX Inc., of Princeton, New Jersey, reported that it has developed a line of transgenic pigs that produce human hemoglobin. But they are still a long way from having hog farmers raise pigs to be human blood donors (Moffat, 1991a).

OTHER INNOVATIONS

Other developments in farm animal biotechnology (which do not entail gene transfer) that can have profound social and economic ramifications include the development of cow clones (Schmickly, 1991) and a technique to preselect the sex of offspring (*Federal Register*, January 10, 1991). Scientists are baffled over the fact that some 25 percent of calves produced by cloning are almost twice the normal size at the time of birth and must therefore be delivered by Caesarian section.

While no plant genes have been inserted into animals, animal genes have been successfully incorporated into the genetic structure of various plants. Tobacco plants have been successfully implanted to produce functional human antibodies, which may be used for diagnosing and treating human diseases. The "antifreeze" gene of the flounder that produces a protein to stop the fish from freezing, has been cloned and inserted into tomatoes and tobacco; crops may be protected from frost in the future by fish genes (Moffat, 1991b).

Since fish farming is on the increase, biotechnologists have been busy developing "superfish," by inserting growth hormone genes from humans, cattle, chickens, mice and other fish into a variety of commercially raised fish, such as carp, rainbow trout, catfish, Atlantic salmon, walleye and northern pike. The antifreeze gene of the winter flounder is also being inserted into other fish species to expand commercial fish production in cold regions and seasons (*New York Times*, November 27, 1990; Manci, 1989; *Biotechnology and Development Monitor*, 1991; Fischetti, 1991).

A biotechnologist at the Army Research Laboratory in Natick, Massachusetts, has cloned the silk–producing gene of the Golden Orb weaver spider and spliced it into bacteria that in turn produce large quantities of spider silk protein. Stronger than silkworm silk and even steel, this new product may have wide commercial use, especially to develop new fabric for bullet proof vests, helmets, parachute cords and other strong, light equipment (AP news release, February 27, 1990).

On the brave new world frontier of medicine, scientists have created a variety of transgenic mice. Some thirty or more strains of mice have been created that develop various kinds of cancers that affect the mammary glands, pancreas, liver, stomach, bones, brain, eyes and kidneys (Adams and Cory, 1992). Another line of mice have been created that carry human genes that result in deformed red blood cells, providing a new model for sickle–cell anemia (*Genetic Engineering News*, June, 1991), and a line of rats have been developed that carry the defective human gene HLA–B27 that causes a painfully crippling form of arthritis (Fackelmann, 1990). The clinical relevance of these new creations has yet to be demonstrated. Making them transgenic provides no foreseeable benefit to the animals themselves, except perhaps for endangered and genetically "fragile" or defective species, like the cheetah and South American maned wolf.

Research is continuing on the identification of genes responsible for various inherited diseases, especially in purebred dogs and livestock and on genes that play a role in development, growth, milk and egg production, disease resistance and other physiological processes. The results of such costly research may eventually be of benefit to animals in terms of their health and overall well–being. But the benefits will be limited if this approach becomes overly reductionistic and utilitarian and is not integrated with a more holistic, if not traditional, approach to improving animal health and well–being. And especially if it is focused primarily on enhancing the exploitative value of animals.

The human genome is being sequenced and genetic defects and strengths identified. Next will be the cow, the pig and the dog. All to what end? New medical and veterinary products and services will certainly result, including varieties of more productive and disease–resistant livestock. But genetic determinism can lead ultimately to eugenics. And eugenics means genetic imperialism and a new world order for a New Creation. Do we really want or need a Creation made over into a human image of perfect utility?

NEW ANIMAL DRUGS

One potential benefit of biotechnology to animals is in the development of genetically engineered vaccines (including birth–control vaccines), hormones, immune system enhancers and diagnostic and screening tests. However, this new generation of veterinary products and services may be a mixed blessing. It is not without potentially adverse animal health, socioeconomic and ecological consequences–as with BST or bovine growth hormone (Wheale and McNally, 1990; Gendel, 1990). Many of these products are no substitute for humane animal husbandry, sound breeding and good nutrition. There is also some evidence that genetically engineered, modified live virus vaccines may play a role in the development of autoimmune diseases, especially in purebred dogs (Dodds, 1990).

PUBLIC ATTITUDES

While private industry and government–funded research centers push forward to create genetically engineered animals that may prove profitable to agribusiness and the medical–industrial complex, the public's attitude toward such developments is noteworthy. In a recent poll across Europe:

> fewer than half thought biotechnological research on farm animals "to make them resistant to disease, or grow faster" should be encouraged. A third thought applying biotechnology to animals "to develop life–saving drugs or study human diseases" was morally acceptable, "provided the animals' welfare is safeguarded," but 20 percent said it was morally wrong and 27 percent said government should decide each case. Only 13 percent thought such work justified "some animal suffering" (Mackenzie, 1991).

A national survey in Japan revealed that 67 percent of people polled were opposed to research that could lead to new forms of plant or animal life (Holden, 1988). Opinion polls in the U.S. show that in 1985, 34 percent of the attentive (informed) public wished to prohibit the creation of new forms of animal and plant life (Feinstein and Miller, 1991). A recent survey in The Netherlands finds that consumers "are very unhappy about eating meat from genetically engineered animals. They are either afraid it will harm them or worried about it on ethical grounds" (Coghan, 1991).

ANIMAL PATENTING

The controversy over the patenting of genetically engineered animals began after the U.S. Patent and Trademark Office ruled on April 7, 1987, that such animals, provided they were nonnaturally occurring "manufactures" and "compositions of matter" could be included under Section 101 of the Patent

Act as patentable subject matter. The patenting of animals was vigorously opposed by The Humane Society of the United States (HSUS) and a coalition of concerned organizations. On August 5, 1987, Representative Charles Rose introduced legislation (HR 3119) to impose a moratorium on the patenting of animals so that the potential adverse implications of such patenting could be carefully studied. And on February 29, 1988, Senator Mark Hatfield introduced moratorium bill S 2111 in the Senate. But on April 13, 1988, the U.S. Patent Office and Trademark Office issued Patent Number 4,736,866 on Harvard University's, DuPont Chemical Co. funded, new creation—the "Onco Mouse," a genetically engineered, cancer–prone mouse (Hubbard and Krimsky, 1991).

Since this time, there have been no further animal patents awarded, even though the U.S. government and U.S.–based multinational corporations have been pushing for changes in European patent law that currently prohibits the patenting of animals (Watts, 1991); and even though the State Department effectively squashed the Rose and Hatfield bills on the grounds that they would weaken U.S. economic competitiveness in the world marketplace.

Some 145 patent applications are now awaiting approval at the U.S. Patent and Trademark Office. Approximately 80 percent of these have medical utility while the remainder involve agricultural animals (*Congressional Record–Senate*, June 13, 1991). One explanation for the fact that no new animal patents have been awarded is that there is as yet no clear regulatory structure set up for the commercial marketing of transgenic animals (Charles, 1991; Fox, 1991).

A new bill was introduced in the Senate (S 1291) by Senator Hatfield on June 13, 1991, to impose a 5–year moratorium on the granting of patents on invertebrate and vertebrate animals including those that have been genetically engineered. HSUS supported this bill with the following statement published in the *Congressional Record* on that day:

> In order for society to reap the full benefits of advances in genetic engineering biotechnology, the social, economic, environmental and ethical ramifications and consequences of such advances need to be fully assessed. Considering the rapid pace of developments in this field, which will be spurred on by the granting of patents on genetically altered animals, a 5–year moratorium on the granting of such patents is a wise and necessary decision. A moratorium will enable Congress to fully assess, consider and respond to the economic, environmental and ethical issues raised by the patenting of such animals and in the process, establish the United States as the world leader in the safe, appropriate and ethical applications of genetic engineering biotechnology for the benefit of society and for generations to come (pp. 7818–19).

It is very likely that the Council on Competitiveness, chaired by Vice President Dan Quayle, will attempt to block this bill. This same Council has been actively working to deregulate the entire biotechnology industry. Its proposed administrative and regulatory guidelines for the Environmental Protection Agency and U.S. Department of Agriculture are such that the risks and costs of new biotechnologies—socially, economically, environmentally and in terms of animal–welfare—will be virtually ignored (Charles, 1991; Fox, 1991).

Clearly, while the genetic engineering of animals is not likely to be stopped, increasing public awareness and censure of the biotechnology industry and its political allies is essential. A 5–year moratorium on the patenting of "new" animal creations would be prudent and timely, especially since we are moving into a new world order of free trade, which should be conditional upon effective international regulations and the adoption of the most stringent controls and regulations over biotechnology by all nations. Otherwise, the privatization of the world's resources and of the genetic material of life itself, coupled with the misapplication of genetic engineering biotechnology in agriculture and medicine, will be against the public interest and the good of generations to come.

CONCLUSION

There are several interrelated dimensions to fully evaluating the costs and consequences, risk and benefits of new developments in science, technology and industry, especially in genetic engineering biotechnology; and of the patenting of both processes and products. These dimensions are as follows: ethical and religious, legal and political, social and economic, environmental and cultural. Generally these dimensions of concern, constraint and direction have been virtually ignored by policy–makers and even seen as obstacles to economic growth and industrial expansion. As a consequence, the gap has widened between private (corporate) and public interests. We are witness to a widening of this gap with the rise of a global industrial biotechnocracy, the costs and consequences, risks and benefits of which need to be rigorously evaluated. To voice such concern should not be misjudged as anti–science, anti–progress. Rather, it should be recognized that with greater involvement of an informed public in the policy decision–making process, advances in science and technology and in biotechnology in particular, will be more likely to serve the public good and to help enhance the quality of life and environment alike. Current attempts by the U.S. government to deregulate the biotechnology industry (Phelps, 1991) and by the EEC's Commission on Biotechnology to eliminate socio–economic considerations in the licensing of new animal drugs (Phelps, 1991) support the conclusion that the direction being taken by the biotechnocracy of the industrialized world is neither prudent nor appropriate.

Some proponents of genetic engineering feel "more comfortable" (from a not fully articulated moral/ethical perspective) with the patenting of the *techniques* of biotechnology, rather than with its products, including transgenic animals. However, the unconditional acceptance of creating transgenic animals for any and all purposes (from the perspective that it is not a moral/ ethical issue to create such animals) is as unreasonable as the unrealistic abolitionist position that would prohibit all such research and development (because it is immoral/unethical).

Our power over the genes of life is a recent acquisition, as significant a notch in human evolution as the discovery of pyrotechnology and atomic energy. But this power does not give U.S. the absolute right to transform animals to further satisfy our myriad needs and wants. Rather, it places U.S. on a critical threshold of moral/ethical choice and responsibility. This means that we must choose wisely and compassionately, case by case. We must not forget the history of science, technology and industry. In the past, we have made many wrong choices for selfish reasons, the consequences of which have been as harmful to our own kind as to the rest of Creation, especially to the animal kingdom and to the natural biodiversity of our fragile planetary ecosystem.

ADDENDUM

First Creation–First: Protecting the First Creation from Further Desecration and Transformation

The kinds of plants and animals that are being genetically engineered for agriculture (along with a host of other agribiotechnology products) are primarily those kinds that are being designed for adoption by conventional agriculture. Their adoption and incorporation into our food production system should be contingent upon them quickly helping make industrial agriculture humane and sustainable.

The appropriate use of agribiotechnologies in ecological farming, in holistic resource management and in the development of alternative, socially just agriculture is possible and attainable. It should not be used as another technological fix to compensate for the effects of agricultural degradation, but at the same time aggravate that degradation, necessitating even more costly "fixes." It is surely absurd to use this technology to boost productivity of agricultural commodities that are already in oversupply, like milk.

Genetically engineered bovine growth hormone (BGH)—which is an affront to the science and ethics of good dairy cow husbandry—is the first product that the biotechnology industry has yet to recognize as their own Ford Edsel: And selling it to good dairy farmers is like convincing Eskimos that they need refrigerators.

Increased dependence upon biotechnology will put us on the treadmill of economic competitiveness accelerating the transformation of life into profitable commodities, with the emergence of genetic imperialism and an increasingly parasitic relationship with the rest of Earth's creation, as exemplified by turning farm animals into bioreactors to produce pharmaceuticals.

Another example of gross misapplication of biotechnology would be to develop a vaccine to make African cattle resistant to Trypanosomiasis rather than using this new technology to increase overall herd health and productivity and indirectly reduce herd size. Such a vaccine would lead to an expansion of livestock into wildlife areas (where wild animals possess natural immunity) and will mean the end of the wild.

While the benefits to animals of making them transgenic are unclear, there are clear benefits of other biotechnologies to enhancing their overall well–being. These include: rapid identification and elimination of genetic diseases; increased disease resistance; protection of endangered species; humane population control of feral and wild species; preservation of genetic diversity; selecting farm animals better adapted to traditional, and alternative, humane husbandry systems; increased efficiency/productivity of farm animals, which will mean fewer animals, more efficient resource utilization and more land being freed up for wildlife habitat recovery.

Appropriate uses of biotechnology in animals should follow the "3R's" of *refinement, reduction* and *replacement* in the utilization of animals by society today. We need to not only decrease the suffering and enhance the well–being of animals utilized by society today, we also need to decrease and not increase our dependence upon them for a host of reasons—economical, ethical, environmental, etc. For detailed discussion see Fox, 1992.

Gene mapping and marker–assisted selection to identify useful genes in cattle, hogs and poultry should not be focused primarily on making these animals more productive under conventional husbandry conditions. Overproduction is a chronic and unacceptable problem, lowering farmers' profits and forcing them to get bigger or get out. Better to seek genes that will help livestock and poultry be healthier and better adapted to more ecologically sound farming practices, like rotational grazing of dairy cows and pasture feeding of hogs and helping the livestock population in the Third World cease to expand and to become healthier and more productive. Better still, perhaps, to conserve and propagate rare breeds for such purposes than to create patented transgenic animals derived from narrow utility stock genetically selected for generations to be used under intensive factory systems of production that are gaining widespread societal disapproval.

Above all, new developments in biotechnology should not create *barriers* that would prevent or delay the adoption of alternatives, such as more humane sustainable animal husbandry practices and greater advances in public health, education and nutrition instead of creating ever more transgenic mice. With these considerations and caveats, the appropriate application of biotechnology in animals will be more reasonably assured and objectively determined.

The biotechnology explosion has resulted, over the past decade, in the creation of over 10,000 new lines of transgenic mice; in farm animals with human genes producing milk with new health promises to offer genetically impaired and immune–compromised people; in genetically engineered plants secreting spider venom. Human disease antigen injected into cows to provide day care infants and others with protection from the diseases that are spread and potentiated by their situation may soon be marketed. Milk contains a natural opiate, which may help calm some kids down. Selecting cows to produce opiate–rich "Sleepy Time" milk may be on the horizon soon. Already there is a company developing a transgenic pig industry to provide "xenografts"—genetically humanized swine hearts, livers and kidneys—for humans in need of such organ transplants.

These new directions and applications of biotechnology make one wonder when there will be a concerted effort to develop and distribute a safe and effective, if not also a reversible, genetically engineered contraceptive. That it will be developed and marketed for women first should not be an obstacle to its widespread adoption. The Catholic Church could help by embracing the view that such an application of biotechnology is to use our God–given power over the gene for reasons of compassion and to further the greater good and future security of all Creation.

We must stop multiplying our numbers, needs and wants. And we must learn to live gently and simply so that others may simply live. We need to stop regarding technological progress as unstoppable. It is *change* that we cannot stop; and it is up to us to direct progress, to take charge of the direction new technologies might take in order to maximize their benefits and minimize their costs and risks to all concerned. We need to constantly redefine what true progress is in order to implement correctives and preventatives where needed.

The mutant "monster" creation of genetic engineering we all feared has already been created and released into the world. It is not some insulation–eating superbug, AIDS–like virus, mind–altering transgenic pollen, or Iowa corn plant that eats Texas beef. This monster seeks to use biotechnology for purely materialistic and consumptive ends and is preparing to remake creation into its own image of how the natural world can best serve its myriad consumptive needs.

The primary purpose behind the genetic engineering of Second Creation products is their profitability to their makers: the remakers of the First Creation. The mythic image of this mutant monster that connects us to a fate far more terrible than that of a Midas, Icarus, Prometheus or Marsyas, is so because it consumes itself as it destroys the Earth's natural resources.

After the carnage, the pestilence, the long drought, the ozone hole, Chernobyl, pesticide rain and shores reeking with dead seals and dolphins, what will there be? What is coming is what we see, unfolding before our very eyes.

Genetic engineering and all applicable technologies should first and foremost be directed at these kinds of issues and with the vision and ethics of organic, if not sacred, unity rather than at developing new biotechnology products to help boost a non–sustainable agricultural system and an unethical biomedical research industry. The so–called health industry fights cancer with tons of profitable treatment, but not an ounce of prevention and justifies untold animal suffering in the name of medical progress.

How can we have a government with health and agricultural agencies that are not in concert, but in kahoots? For them to prohibit the wholesale application of thousands of potential carcinogenic chemical pesticides and the millions of tons of petrochemical fertilizers by the feed and food industry would be in the public interest. So why has this not been done and countless other social and ethical issues addressed by industrial world governments? Perhaps not until we all confront our own personal monsters and demons and discover that we are all related.

We cannot continue to be blind to the irony that there are many publicly supported corporations that are distributing pesticides, as well as processing and marketing various crops and factory–farmed animals; developing and patenting genetically engineered mice and selling x–ray film for nationwide, annual mammograms.

The monster on this planet is a product of biotechnology because in breaking the DNA code, it became addicted to changing the codes of life to serve its own industriously Earth–transforming and all–consuming existence.

The male of this monstrous product of biotechnological skill and arrogance likes those of its opposite sex to have large firm breasts. It even sells them breast implants when their natural breasts become cancerous with the poisons of industrial indifference, ignorance and greed.

The entire immune system within the monster body–human, the *corpus* of industrial civilization, is beginning to break down. It is creating thousands of varieties of transgenic mice to find ways to help its failing immune system adapt to an increasingly dysfunctional and hazardous society and environment. The monster's name is Nemesis. His mate, more ignorant than Eve and not so loving and alive, is a modern Pandora with perfect silicone breasts.

Where nature is the least defiled, humans still go to rest, dream, heal, play and pray. But some still come to these sacred places—the dying remnants of the First Creation—simply to take. They must be stopped by all and every means for our children's sake and for all other creatures great and small.

To conclude, from an ethical and spiritual perspective, the future of the natural world or First Creation will only be secure if this new technology is applied like no other before it. Otherwise, the Second Creation will mean a wholly unnatural, humanized world as we transform the First Creation into a bioindustrial system remade into our own self–serving materialistic image of utility and productive efficiency.

REFERENCES

Adams, J.M. and S. Cory. 1992. Transgenic models of tumor development. *Science*. 254:1161–64.

AP News Release. 1990. Army's gene–spliced spider silk may prove superfabric of future. Boston. February 27.

Bialy, H. 1991. Transgenic pharming comes of age. *Bio/Technology.* 9:786–88.

Biotechnology and Development Monitor. 1991. Biotechnology in fishfarms: Integrated farming or transgenic fish? 7:3-6.

Charles, D. 1991. White House changes rules for genetic engineering. *New Scientist,* May 25, p. 14.

Clark, A.J., P. Simons, I. Wilmut and R. Lathe. 1987. Pharmaceuticals from transgenic livestock. *Tib. Tech.* 5:20–24.

Coghan, A. 1991. Dutch lack appetite for genetically 'altered' foods. *New Scientist.* August 17, p.9.

Congressional Record–Senate. 1991. Letter to Hon. Mark O. Hatfield from H.F. Manbeck, Jr., Assistant Secretary and Commissioner of Patents and Trademarks. June 13, p. S7817.

Durning, A. and H. Brough. 1991. *Taking stock: Animal farming and the environment.* World Watch Institute, Washington, DC.

Dodds, W.J. 1990. Vaccine, drug and chemical–mediated immune reactions in purebreds challenging researchers. *DVM Magazine.* December, pp. 41–42.

Fackelmann, K.A. 1990. Engineered rats reveal arthritic surprise. *Science News.* December, p. 357.

Federal Register. 1991. Agricultural research service intent to grant and exclusive license to animal biotechnology. January 10, 56 (7):990.

Feinstein, P. and J.D. Miller. 1989. Public perception of biotechnology: Is the glamor gone? In *Biotechnology Seminar Series Academic Year 1988–1989: Summary Reports.* Tufts Center for Animals and Public Policy, Tufts School of Veterinary Medicine, North Grafton, MA. pp. 12–14.

Fischetti, M. 1991. Feast of gene–splicing down on the fish farm. *Science.* 253: 512–13.

Ford, J. 1988. This little pig rushed to market. *New Scientist.* April 28, p. 27.

Fox, J.L.1991. Scope proposal goes another round. *Bio/Technology.* 9:603.

Fox, M.W. 1992. *Superpigs and Wondercorn.* Lyons and Burford, NY.

Fox, M.W. 1990. The cattle threat. *The Humane Society of the United States News.* Spring, pp. 24–27.

Gendel, S.M. and A.D. Kline, D.M. Warren and F. Yates, eds. 1990. *Agricultural bioethics: Implications of agricultural biotechnology.* Iowa State University Press, Ames, Iowa.

Genetic Engineering News. 1991. Transgenic mice development for sickle–cell anemia. June, p. 34.

Holden, C. 1988. Japanese views on science compared to U.S. attitudes. *Science.* 240:277.

Holden, C. 1991. Superchicken. *Science.* 253:265.

Hubbard, R. and S. Krimsky. 1988. The patented mouse. *GeneWatch.* 5:6–7.

Kraemer, D.C. and J. W. Templeton. 1990. Genetically engineered resistance in mammals. In *Veterinary Perspectives on Genetically Engineered Ani mals.* American Veterinary Medical Association. Schaumburg, Illinois. pp. 48–53.

Mackenzie, D. 1991. People's poll shows confusion over biotechnology, *New Scientist.* July 13, p.14.

Manci, W.E. 1989. Researchers continue work in transgenic catfish and salmon. *Ag. Biotech. News.* Sept/Oct. p. 22.

Mercier, J. C. 1987. Genetic engineering applied to milk–producing animals: Some expectations. In Commission of the European Communities Animal Husbandry Research Program, *Exploiting New Technologies in Animal Breeding.* Oxford Univ. Press, Oxford. pp. 122-31.

Moffat, A.S. 1991a. Three little pigs and the hunt for blood substitutes. *Science.* 253: 33.

Moffat, A.S. 1991b. Bumper transgenic plant crop. *Science.* 253:33.

Murray, J.D. and Rexroad, C.E. Jr. 1991. The development of sheep expressing growth promoting transgenes. In *NABC Report 3 Agricultural Biotechnology at the Crossroads.* J. Fessenden MacDonald, ed. National Agricultural Biotechnology Council. Ithaca, NY. pp. 251-263.

New York Times. 1990. New prospects for gene–altered fish raise hopes and alarm. November 27, p. C4.

New Scientist. 1992. Australian sheep let their hair down. January 4, p 8.

Phelps, A. 1989. Researchers from the Netherlands design cow for production of human–like milk. *Feedstuffs.* September 4, 81:37.

Phelps, A. 1991. EC plans to end socio–economic animal drug criteria. *Feedstuffs.* July 15, p. 25.

Schanbacher, F.L. 1990. Molecular farming: Current status and prospects. In *Veterinary Perspectives on Genetically Engineered Animals.* American Veterinary Medical Association. Schaumburg, Illinois. pp. 54–57.

Scherer, R. 1991. Peelable wool not shear fantasy. *The Christian Science Monitor.* April 17, p. 12.

Schmickly, S. 1991. Don't have just any cow–Clone a better bossy. *The Washington Times.* April 19, p. B7.

Slater, D.W. 1990. Genetically engineered disease resistance in poultry. In *Veterinary Perspectives on Genetically Engineered Animals.* American Veterinary Medical Association. Schaumburg, Illinois. pp. 44–47.

Watts, S. 1990. Drugs industry turns animals into 'bioreactors'. *New Scientist.* April 14, p. 26.

Watts, S. 1991. A matter of life and patents. *New Scientist.* January 12, pp. 56–61.

Wheale, P. and R. McNally, eds. 1990. *The Bio–Revolution: Cornucopia or Pandora's Box?* Pluto Press, London and Winchester, Massachusetts.

Dorothy Nelkin
Sociology and
the Faculty of Law
New York University

*Living Inventions: Biotechnology and the Public**

This presentation will provide an overview of the broad set of public concerns about animal biotechnology that have been expressed both in the U.S. and in the European Community (EC). I will also assess the meaning of biotechnology disputes for recent policy decisions concerning regulation.

The Bush Administration policy on regulating biotechnology defined and limited the scope of statutory authority of federal agencies. To facilitate federal approval of new products and, thereby, to enhance the competitiveness of the American biotechnology industry, the Council on Competitiveness in the Office of the Vice President limited regulatory authority to the issue of "reasonably foreseeable risk to health or the environment." The Council's policies for biotechnology for the 1990s would focus on encouraging economic competitiveness over other concerns.

Limiting regulation to narrow questions of risk was welcomed by some for its economic benefits, but it was promptly attacked by others for ignoring the troublesome implications of creating genetically engineered organisms. These implications have been the source of continuing disputes over issues extending far beyond the question of "reasonably foreseeable risk." The range of these issues suggests that the narrowly focused Bush administration policy and its arguments for the urgency of economic competitiveness are unlikely to ease the growing tensions over biotechnology developments and, indeed, may increase the "intuitive mistrust" that has long marked public attitudes towards genetic manipulation.

THE PUBLIC ANIMAL PATENTING DISPUTES

The patenting of living organisms has become an important focus of these tensions in both the U.S. and in Western Europe. A 1987 decision by the U.S. Patent and Trademark Office held that animals altered by genetic engineering were patentable. However, the European Patent Office decided, in 1989, that it could not grant patents on animals under the terms of the European

* Developed from her paper: *Living Inventions: Animal Patents in the United States and Western Europe.* Stanford Law and Policy Review Vol IV. 1992.

Patent Convention. Currently the issue is the focus of intense debate in several European countries. It is also the subject of debate in the European Parliament as it responds to a legal appeal against the patenting of a genetically engineered human hormone called Relaxid, and considers legislation on the legal protection of biotechnological inventions.

Normally business in a patent office is viewed as technical and hardly a subject for public debate, but the granting of patents to living inventions has spurred a storm of protest from a broad range of interests. The outraged response reveals the complex set of concerns that has more broadly marked public attitudes towards genetic engineering and raises questions about the viability of limiting the scope of regulation.

The debate over the patenting of genetically engineered animals offers a window on the range of concerns that have followed the diffusion of new biotechnology developments since the 1970s when recombinant DNA research evoked fears of the escape of lethal organisms into the environment. Subsequently, critics have mobilized to oppose the field testing of genetically engineered bacteria designed to inhibit frost damage, the creation of genetically altered fish, the development of disease–resistant crops, the use of bovine growth hormones and many other biotechnology applications. Such disputes are likely to amplify in response to advances in human genetics, especially as scientists seek to patent human genetic material.

To explore the nature and diversity of these concerns, I find it useful to present the views expressed by participants in a Congressional hearing and by the literature disseminated by several European Green groups. From there, I am able to provide suggestions regarding the kinds of strategies that may help to develop acceptable policies concerning a technology with significant economic and ethical implications.

Disputes over genetic manipulation of animals have been smoldering in both the U.S. and Western Europe since the 1980 Supreme Court ruling on patenting living organisms. Researchers in both academia and industry have maintained that patent protection of transgenic animals is essential for the development and diffusion of promising medical and agricultural benefits to society and, in its 1987 decision, the U.S. Patent and Trademark Office concurred. After this decision, applications flooded the Office, but so too did protests—from farmers, religious leaders, environmentalists and animal rights activists who were concerned about the consequences of this technology. Farmers believe that the high cost of raising, breeding and owning genetically altered livestock would cause small farm foreclosures. Others believe that patenting animals is immoral because it defines complex, living organisms as profit–making machines. Still others worry that the institutional collaborations between industry and universities involved in the development of biotechnology will compromise the quality and integrity of research.

These issues were played out in the U.S. at Congressional hearings and in Europe in appeals against the European Economic Community (EEC) directive on biological patents. These animal patent disputes aired a set of controversial questions: Would the effects of biotechnology on agricultural production destroy small farms unable to pay royalties? What does patenting imply for the integrity of species and the moral obligation to preserve nature? What would be the effect of patenting on scientific research? The decision to patent transgenic animals has generated economic speculations about the current pace of research in internationally competitive scientific fields and the impact of new technology on the traditional farm sector. Biotechnology and pharmaceutical firms regard patent protection as essential to fueling invention by the private sector, particularly because federal support for agricultural research has declined steadily since the second World War.

American corporate interests, for example, view patenting as necessary if the U.S. is to effectively compete worldwide for the products emerging from biotechnology research. In the Congressional hearings of 1988, Richard Godown, president of the Industrial Biotechnology Association (IBA) pointed out that: "A Japanese company has genetically engineered silkworms to produce a hepatitis vaccine" and "the United Kingdom and Ireland may be in the lead in animal biotechnology." Even China, he observed, is already test marketing low–fat pigs produced by growth hormone injection. Another scientist asked the Congress: "How are our farmers going to feel when that ham, which is 70 percent fat–free, comes here in cans and is sold in the United States?"

Farmers, however, are deeply divided on the issue. The American Farm Bureau Federation, representing 3.5 million member families, has long favored free market policies and innovation in farming technology, and emphasizes the potential benefits of genetic engineering. They anticipate that new biotechnologies will reduce farm costs and expand the utilization of farm products—transgenic livestock would grow faster and be brought to market sooner than conventional breeds; genetically lean and disease–resistant animals would provide healthier meat to consumers than the fatty livestock injected with antibiotics sold today. According to the Farm Bureau, patents and exclusive licensing to the private sector were the only ways to ensure the development of biotechnology discoveries and the commercialization of new agricultural products.

Other farmers' groups, however, oppose animal patents on the grounds that large corporations would usurp the ownership of livestock—a resource presently controlled by the farmers themselves. These farmers predict dire consequences for traditional breeding techniques and they worry about the continued demise of the small family farm. "Will animal patenting result in greater concentration in those who produce breeding stock as it has created in the seed industry in recent years? ...We are an industry that has seen our numbers drop drastically in the past two years and there are not enough of us left to take chances on a major mistake," said the president of the National Farmers Union (NFU) in his testimony at the hearings. Representing about 300,000 farming families, this organization believes that "The small business structure of the family farm is the nation's bulwark against communism and fascism. It is essential to the democratic way of life." This has frequently translated into opposition to technological changes that would harm the small farmer.

The National Farmers Organization, the American Agriculture Movement, the Coalition to Save the Family and the League of Rural Voters joined the NFU in its concerns about the economic implications of genetic engineering. They played on the popular myth of the family farm as a foundation of American values and the fear that patents would force further corporate concentration of agriculture. European groups, for example, the UK's Compassion in World Farming, oppose patenting for similar reasons. "The losers would be the smaller plant and animal breeders who are not able to embark on research or pay the royalties....This makes the farmer more dependent on the chemical industry that currently controls biotechnology." It would further divide the farming community into winners (those who can afford to adopt these expensive technologies) and losers (those who cannot). In Europe, as in the U.S., opposition focused on issues of equity, drawing from the growing concern about monopolistic practices in the agribusiness, pharmaceutical and chemical sectors.

While farm interests have focused on the economic consequences of animal patenting, an unlikely coalition of religious, environmental and animal rights groups have raised a set of moral concerns. These groups attack the issue from different perspectives reflecting their own moral agendas, but they all reject the definition of animals as resources, or, in the language of patent law, "compositions of matter." Jeremy Rifkin, a persistent critic of biotechnology research, has accused the U.S. Patent and Trademark Office of reducing "the entire animal kingdom of this planet to the lowly status of a commercial commodity—a technological product indistinguishable from electric toasters, automobiles, tennis balls, or any other patented product." While most visible, Rifkin is not alone; indeed, his influence as a biotechnology "gadfly" reflects the wide appeal of his ideas. Representatives from an array of citizen's groups are questioning the moral authority of scientists to alter the state of nature.

Although humans have owned and used animals for millennia, the idea of patented ownership has invited renewed scrutiny of the human–animal relationship. Animal rights group in the U.S. and in Western Europe are enjoying extraordinary expansion and public visibility. They maintain that tinkering with an animal's genes violated "species integrity" and, in typical animal rights language, "the inherent sanctity of every unique being and the recognition of the ecological and spiritual interconnectedness of life." Expressing these assumptions at the Congressional hearing, John Barnes, a veterinarian testifying for the Alliance for Animals, suggested that animal ownership has traditionally implied the responsibility of stewardship. This responsibility would be lost with corporate ownership.

Animal rightists usually see themselves in conflict with those who espouse a Judeo–Christian ethic supporting human domination over animals, but on this issue these groups concur. Reverend Wesley Granberg–Michaelson of the National Council of Churches said, "A real shift in how humanity relates to the natural environment is occurring when we face this issue." From his religious perspective: "The Judeo–Christian view says that the Creation is, in essence, held in trust...We have a responsibility to see that its integrity is preserved."

The notion of "species integrity" has raised the question of what is, or is not, natural. Genetic engineering, sanctioned by patents, seems to some a profoundly unnatural act. "We are engineering ourselves away from natural selection into a mechanical selection of traits," said Representative Charles Rose (D–NC), who had introduced legislation for a moratorium on animal patents. Granberg–Michaelson stated this idea in graphic terms: "Cows do not mate with fish. Humans do not mate with pigs. Fireflies do not mate with tobacco plants. These combinations are more that what can be called simply 'natural occurrences.'" Similarly, European activists have objected to the view of "living factories rather than sentient beings." They reject the very principle of patenting of life as reflecting "a highly questionable relationship of Humanity to Nature." It would "undermine any last thread of respect for nature in our already artificialized world...forcing upon us a reductionist and materialistic concept of life." Thus, this dimension of the debate reflects fundamental philosophical differences concerning the essential nature of living beings.

Some concerns about biotechnology have focused more concretely on the effect of patenting on the research agenda and on the use of government supported science. Is it right that private biotechnology companies will profit by building on a base of publicly funded research? Will academic scientists, in dealing closely with industry, be appropriately accountable for their work?

In the U.S. Congressional hearings, the president of the Farmers' Milk Marketing Cooperative, addressed the issue of profit: "...if most of the research and development costs in the production of some super animal have been paid for from public coffers, is it proper to grant a monopoly market position for giant corporations for 17 years? Furthermore, is all of this necessary to promote alleged scientific progress?" In fact, current laws deliberately encourage private exploitation of publicly funded research as the most effective means of diffusion; but this policy continues to confront opposition mainly from groups concerned about the general direction of technological change.

Many scientists favor the growing industry–university collaborations in commercially useful research. They minimize the risks of such ventures based on traditional assumptions about scientific neutrality and the ability of scientists to regulate themselves. They do not believe patenting would distort the research agenda of scientists which is shaped by intellectual interests and controlled through peer review and the values of their scientific disciplines. Others, however, suspect that the profit motive would, indeed, affect research. Jack Doyle of the Environmental Policy Institute, suggested at the hearings that industry–university collaboration in research and development "is worrisome because it blurs the roles of government as regulator and the university as society's natural arbiter and adviser." Such collaboration would disturb the traditional checks and balances on scientific knowledge and its application and shape the direction of future research. Critics doubt the ability of scientists to control the direction and use of their own research. "Allowing patent protection at this time will sever the contact between research and the public interest," said the Wisconsin Farm Unity Alliance. "It will mean that biotech corporations will be able to finance a much more accelerated level of research and development with little concern for the need to build public understanding and support and even less concern about meaningful regulation." This view was especially troubling in Europe where private industry funding of research in public institutions is a recent practice. Opponents fear the privatization of public research and suspect that patenting would lead to restricted information exchange among scientists and further limit public access to scientific information. They feel that only private industry would benefit.

ETHICAL, ECONOMIC AND POLITICAL CONCERNS

The patenting of animals has become a lightning rod for existing ethical, economic and political concerns in both the U.S. and Europe. It takes place when the plight of small farmers is a growing problem and technological changes in the farm community are a polarizing force. The issue has entered the public arena when animal rights groups are questioning the morality of vivisection and arguing against the instrumental values that allowed animals to be used as a resource. The decision also touches on controversial and widely publicized possibilities of commercializing human tissue for fetal research and human body parts for organ transplantation. It feeds existing worries about the effect of proliferating industry–university collaborations in biotechnology with their implications for the values of open scientific communication, professional responsibility and academic freedom.

The decision also resonates with the general uneasiness about genetic research which relates to vague, yet profound, fears of human genetic engineering. One should not underestimate the depth of public feelings about tampering with genes. We have only to look at the long history of popular culture—films and science fiction—that play on the fear of radiation mutation and genetic manipulation to discover its archetypal roots. Recall, for example, the series of classic horror films in the 1950s (e.g., *The Fly, The Wasp*) and their images of mutant monsters resulting from radiation and tampering with genes: ants, wasps, spiders, scorpions mutated into the size of 747s. In Europe, the discourse on genetic engineering is colored by images of Nazi eugenics and human experimentation. These fears contribute to the opposition to genetic engineering and its popular image as technology out–of–control.

The biotechnology debate must be understood in the context of the many other policy controversies over science and technology, for example, over the practices of fetal research and animal experimentation, the teaching of evolution in the schools, the burial of nuclear wastes and the effects of technology on the environment. Such controversies reflect fundamental, and sometimes irreconcilable, values that are not easily resolved. In the case of animal patenting disputes, the small farmer, economically committed to the family farm and ideologically convinced that it is "essential to the democratic way of life," is not likely to be convinced that patenting is beneficial. Biotechnology is not the cause of the decline in family farming and a ban on animal patents would not reverse the trend. For some, this technology has come to symbolize the differential social and economic impacts of technological change. Similarly, arguments about the usefulness of transgenic animals for medicine and research are unlikely to stop the opposition of animal rights crusaders. Driven by anti–instrumental values and beliefs about the sanctity of nature, they are mobilized to oppose all use of animals as tools. They are particularly troubled by techniques of biotechnology that have blurred the boundaries between inert matter and living objects, techniques now recognized in law and reified by the decision to patent animals as living inventions.

Even the scientists who deny effects of commercialized research on the norms and practices of science base their position on fundamental beliefs about academic integrity. Convinced of the moral neutrality of science, they assert the ability of scientists to resist the lures of profit and to effectively regulate themselves. Though academic engagement in the development and diffusion of new technology surely weakens the credibility of the academy as an independent source of assessment, industry–university consortia in biotechnology have proliferated.

The controversy over biotechnology patents has developed out of a fundamental clash of moral values, conflicting visions of progress and competing world views. Based on beliefs about equality and justice and reflecting questions about the meaning of progress, such controversies cannot be resolved by simply assessing risk or claiming the necessity of greater competitiveness. Nor, in democratic societies, can they be simply dismissed, for underlying protest is a troubling mistrust of the authorities responsible for technological development. A national survey conducted by the Office of Technology Assessment has indicated broad mistrust of the government's role in regulating biotechnology. In disputes involving statements concerning potential risks, Americans believe environmental groups over federal agencies by a margin of 63 percent to 26 percent. The picture is similar in Europe where opinion surveys carried out in the 12 member states of the EEC showed that 52 percent of the people trust environmental and consumer organizations "to tell the truth about biotechnology and genetic engineering." Only 20 percent chose public authorities and 6 percent chose industry as trustworthy guides.

The critical questions, then, have to do with authority. Who should be making decisions about a technology with such broad economic, moral and political consequences? How can we develop policies for technology assessment that would include broader concerns about new biotechnology products and processes? The critical challenge in both the U.S. and the EC is how to develop mechanisms for conflict resolution in the face of intuitive mistrust, competing economic visions and philosophical disagreement about the costs and benefits of new technologies and their differential effects.

Since the early 1970s, similar challenges have been expressed in disputes over other technologies that present potential risks to health, environment or social values. Opposition to nuclear power; protests against the siting of airports, toxic waste dumps, chemical plants and other noxious facilities; and fights over the rules and standards regulating pollution, began in the early 1970s—most have persisted for several decades. To resolve these disputes, public agencies in both the U.S. and Western Europe during the 1970s encouraged greater public involvement in technological decisions on the assumption that this would foster public acceptance of technology and enhance the legitimacy of decision–making institutions. There followed a variety of efforts to involve citizens more directly in creating and implementing policies for technological change.

These efforts ranged from broadly participative inquiries to environmental mediation. They included complaint and consultation systems, citizen advisory groups, representation of citizens in review boards and special issue referenda. Some were intended to develop consensus among conflicting scientific groups as a means to advise decision–makers (e.g., science courts); others to educate the public. The process created depended on how the problem of public acceptance was defined. Where lack of public confidence was thought to arise from technical uncertainties (for example, about risk), the goal was to develop a scientific consensus among dissenting groups in order to improve the advice available to decision–makers. Where problems of acceptance were attributed to lack of public understanding, the task became one of public education. Where controversy was defined in terms of alienation and mistrust, more participatory and consultative systems evolved.

The 1970s experiments had mixed success, depending in part on how realistically they defined the source of public opposition to technology policies. However, they helped to avoid the polarization and mistrust that is so evident in biotechnology disputes today. This polarization reflects, in part, the insistent focus on economic competitiveness as the central value overshadowing all other concerns—a focus that necessarily defines public participation as an impediment to technological change. This attitude, however, has only served to exclude issues of public concern and increase public resentment. European groups, for example, feel that "the public is being kept out of the discussion, as if it were merely a technical matter. This must stop! The patenting of life is too important to leave up to a handful of experts and corporate lobbyists."

CONCLUSION

I conclude by extracting some principles from the 1970s struggles to establish effective negotiations for the resolution of technological disputes. We learned from these struggles that:

—Negotiations must deal directly with issues of public concern including questions of ethics and equity as well as economics and risk. Thus, controversial issues must be defined in terms of problems to be solved rather than solutions to be accepted. Proponants of a technology, determined to implement preconceived decisions, try to deal with protest by co–opting public support rather than by expanding choice. Leaving little room for compromise, these attitudes often resulted in the transfer of conflict from public hearings to the courts and sometimes to the streets.

—We learned that effective negotiation requires that participants have a sense of political efficacy and choice over the issues that most concern them. Establishing political efficacy rests on widely distributed knowledge and access to expertise. High quality educational materials should be designed, not to promote the technology, but to open frank discussion and understanding of both benefits and costs. Thus, efforts to enhance the competence (and to avoid manipulation) of journalists is essential, for the media play a significant role in informing the public.

—Finally, developing trust is a long–term process built on evidence of reliability and openness established over time. The emerging field of biotechnology offers opportunities for policy negotiation early in the development of the technology, before significant choices are made. These choices should not, at this early stage, be limited to narrow, short–term questions of risk. The dispute over animal patents suggests that evaluation of the products and processes emerging from biotechnology must be developed with an eye to their differential social and economic impacts. The institutional procedures for assessing these impacts must also involve those who are affected and concerned. The history of participatory procedures suggests this may not produce consensus; when technologies embody highly controversial political and social values, consensus is not a feasible goal. By sorting out conflicting values, they may reduce public mistrust of administrative institutions and, in the long run, encourage the development of equitable decisions.

PART III OPPORTUNITIES AND CHALLENGES: 4 WORKSHOPS

BIOTECHNOLOGY AND ANIMAL WELL-BEING

David Meeker
Director, Research and Education
National Pork Producers Council

Animal Well–Being and Biotechnology

This topic brings together two issues which have been much on the minds of livestock producers in recent years. Animal welfare and biotechnology are important, not only because the politics of the issues could affect agriculture, but also because producers are good citizens concerned about doing what is right.

A workable definition of biotechnology can be derived by examining the two parts of the word. "Bio" stands for biology, the science of life that includes all living things. "Technology" is collectively tools and techniques which include animal breeding, embryo transfer, genetic engineering, fermentation, tissue culture and so forth. Biotechnology is applying these tools to living organisms to get them to do what you want them to (Witt, 1990).

Biotechnology offers the potential of incredible benefits for society with very little risk, such as a whole generation of safer, more effective drugs. Hundreds are in development including 50 new cancer drugs and 15 new AIDS drugs now being clinically tested (Gorner, 1992). Nearly 8,000 commercial processes which use genetic engineering principles are in the process of being patented. Disease–resistant crops and livestock, more efficient food production, lower fat meat and biotechnology–aided processes can help make significant gains in feeding the world higher quality food. Society has an obligation to develop these techniques.

Biotechnology is playing an increasing role in nearly all scientific fields. To choose not to implement these tools in any one industry or country would leave that industry or country noncompetitive. The application of biotechnology to agriculture has lagged behind human health applications due to a lack of investment which would yield needed basic knowledge in animal physiology, biochemistry and microbiology (National Agricultural Research and Extension Users Advisory Board, 1990).

With all of the intense research efforts to date, not one industrial accident or disaster has befallen society because of biotechnology. This is not to say that food safety, environmental protection and animal welfare issues should not be addressed. They should be addressed, but in an appropriate perspective.

REGULATION

The Bush administration announced in 1992 that no special regulations were needed for gene therapies and genetically engineered drugs and pesticides. This was good news for the biotechnology industry and for society. The same logic should be applied to the issue of animal well–being in biotechnology.

Biotechnology should be considered as one more in a long line of tools developed for the betterment of human life. Once human and environmental safety are proven for biotechnological procedures and products, their use should be allowed. Additional regulations pertaining to animal well–being in a society utilizing biotechnology are not needed.

It is human nature to develop a system of ideals, practices and prohibitions to both protect us from nature and from ourselves (Kaye, 1992). There are many reasons for society to have regulations and to vigorously enforce them. However, regulations are not always the best way to affect human behavior. Regulatory activity should be focused on the priorities of protecting and enhancing human life. It is not practical, possible, or cost–effective to regulate every aspect of industry and research.

Additional regulations on industries and people who are willing and capable of doing what is right are a waste of time and effort. Unnecessary regulations stifle competitiveness by burdening industries with unproductive paperwork, delays and bureaucracy. The key is to give people involved in production and experimentation the training and information to act responsibly. A more humane, enlightened and compassionate regard for all life, including human life, is a mindset that cannot effectively be legislated. People continue to be bound by a moral obligation to minimize pain and suffering of animals while advancing important interests of their fellow human beings.

ANIMAL RIGHTS AND ANIMAL WELFARE: AN OVERVIEW

Philosophical conflicts about whether or not animals have rights dates back to early civilization. The Greek scholar, Pythagoras, was a vegetarian who believed human souls migrate to animals after death. On the other side was Aristotle, who believed that animals existed to serve humanity. The 17th Century French scientist, Descartes, believed that humans alone have souls and on this basis he categorized humans separately from all other matter. He considered animals as machines with no capacity for pain.

Judeo–Christian traditions and teachings support the concept that animals and humans do not have similar interests or rights. Old Testament writings described humans as having dominion over all creatures. The use of animals was permitted for food, service, protection and even sacrifice. New Testament concepts generally support Old Testament descriptions of humanity's dominion over animals, but stopped the practice of animal sacrifice. Some biblical passages encourage kindness to animals. Of course, other religions, such as Buddhism, have very different perspectives on human–animal relationships.

Howard Kaye (1992) wrote, in an essay lamenting the reductive and deterministic view of human life accompanying the Human Genome Project, that more than the categories of heredity and environment are required for understanding human life. He said that humans are moral and cultural beings with the elements of will, choice and responsibility contributing to the essence of their being. He wrote, "Our capacities for reason, symbolic expression and imagination; our aspirations for esteem and respect; and our qualities of curiosity and self–consciousness all may have evolutionary origins and may have contributed to our species' biological success." While Kaye uses these arguments to say that humans should be seen in much more than just a biological sense, it also follows that human beings are, in many respects, very different than animals.

While most people and most farmers believe that animals have no rights, they do believe that animals (and humans) should be spared unnecessary suffering through neglect, deprivation or willful abuse. There is a great difference between the humane treatment of animals and humanizing animals. This important difference between animal welfare and animal rights seems to be inherent in the thinking of most people. But as society becomes more affluent and more well–fed, there are more resources available for social movements and philosophical thinking about topics such as humans' relationships with animals. That society dwells on questions of animal rights and animal welfare when problems of human welfare, human rights and world hunger abound should make us stop and examine our priorities. The truth is that farm animals are treated reasonably well and that the use of animals in medical research greatly benefits people. The use of biotechnology does not change the basic responsibilities that humans already have toward animals as they farm or do research.

ANIMAL RIGHTS AND ANIMAL WELFARE: A CLOSER LOOK

Rare are discussions of animal rights/animal welfare and the use of animals by humans which are objective, scientific and cogent. Two writings which are particularly useful in sorting out the complexities are: *The Ethics of Meat Production* by Kauffman and Rutgers (1991) and *Interspecific Justice* by Van de Veer (1979).

Kauffman and Rutgers reason that animals do not have rights since they cannot exercise or respond to moral claims. Beings with rights must balance their own interests with what is just. Therefore, humans have a moral obligation to treat animals with compassion, prohibit cruelty, prevent extinction of species and respect animals' basic interests. Basic interests of animals include freedom from pain and suffering, nourishment, freedom of movement, companionship of other animals and protection from predators. The "five freedoms" of the UK guidelines for animal welfare in agriculture could also be taken to define farm animals' basic interests.

Kauffman and Rutgers admit that most professionals in meat, animal and veterinary sciences have not taken time to thoroughly examine the moral justifications or ethical decision–making on these philosophical issues. They write in detail about the definitions of philosophy, ethics and rights. Ethical principles are presented as guides in making decisions and exercising judgments about how we think about and treat humans and animals. These authors emphasize that before animal–human relationship issues can be resolved responsibly, people must think through all the major issues. Each person should make his or her own decision independently before collective societal decisions can be established. They conclude that the use of animals for food or for experimentation to provide for humans' right to live healthy lives is justified, but that the well–being of those animals should not be ignored.

Van de Veer (1979) cites five ways that humans can relate ethical principles to animals:

—*Radical Speciesism (RS)* is the extreme view held by Descartes, but few people presently believe that animals are objects having no interests. This view would allow people to use animals in any conceivable way without any regard for animal well–being.

—*Extreme Speciesism (ES)* maintains that an animal does have certain interests and needs and is more than an object at the disposal of man. ES would, however, permit subordination of basic interests of animals for even peripheral interests of humans. Most people would reject ES as well as RS because it would allow animal suffering as long as some peripheral human interest was being served.

—*Interest Sensitive Speciesism (ISS)* is the view that when there is a conflict of interests between an animal and a human being, it is morally permissible to subordinate animal interests to promote basic interests of humans. However, one may not subordinate basic interests of animals in promoting peripheral human interests. ISS would permit the sacrifice of a dog to save a human life, but would not permit animal suffering for frivolous reasons. A majority of people would subscribe to this ethical principle, though classifying various interests as basic or peripheral could be a problem, especially among species at different stages of the evolutionary ladder.

—*Species Egalitarianism (SE)* gives animals equal status to humans when interests are considered. It holds that when there is a conflict of interests between an animal and a human it is morally permissible to subordinate the more peripheral to the more basic interest and not otherwise, regardless of which one is jeopardized. Few people would subscribe to SE.

—*Two–Factor Egalitarianism (TFE)* holds that interests and psychological capacities are both important factors as conflicts of interests between two beings are considered. Many people would subscribe to TFE along with ISS. TFE would allow the sacrifice of an interest of a species with less developed psychological capabilities to promote a like interest of a more developed species. Basic interests of the lower species could be sacrificed for promotion of serious interests of the higher species. TFE attempts to take into account both the kind of interests at stake and the psychological traits of the beings in question. As in ISS, there is difficulty in objectively assessing these interests and capacities.

Since most people would subscribe to the ISS and TFE approaches described by Van de Veer, humans must consider animals' interests and psychological capacities. Research should continue on animals' perception of pain, stress quantification, healthy physiology, behavioral characterization and the interaction of productivity with these factors.

The degree of morally acceptable animal suffering is higher for medical experimentation than it is for animal agriculture. Biotechnology can likely produce better animal models to study human disease. The potential for great human benefit from these genetically engineered animal models outweighs the fact that animal well–being may be decreased. Biotechnological advances such as genetic engineering may also make it possible to increase the well–being of disease–model animals by making them more able to cope with their surroundings, less susceptible to stress and less sensitive to pain.

EUROPEAN ANIMAL PROTECTION

In 1964, Ruth Harrison's book *Animal Machines* was published. Consequently, the UK government set up a Technical Committee to examine animal welfare in intensive livestock systems. The 1965 report of the Brambell Committee led to the establishment of legal definitions of behavioral needs. The Farm Animal Welfare Committee of the UK has articulated criteria to assess animal welfare in agriculture. These so called "five freedoms" are:

1. freedom from thirst and hunger;
2. comfort and shelter;
3. prevention/rapid treatment of disease;
4. freedom to display most normal patterns of behavior;
5. freedom from fear.

In 1976, a European Convention for the Protection of Animals kept for farming purposes was elaborated by an *ad hoc* committee comprising delegations from most of the member states of the Council of Europe. Ingvar Ekesbo (1991) urged this convention to include some basic rules that limit humans' right to manipulate animals kept for farming purposes. However, he adds that such rules should not limit the possibility to do research in biotechnology. The rules suggested by Ekesbo regarding biotechnology are:

—Animals produced as a result of genetic manipulation procedures shall not be kept for farming purposes unless, through scientific evidence, it is shown that their health and welfare will not suffer;

—No substance shall be administered to an animal kept for farming purposes unless it has been demonstrated by scientific studies of animal welfare that the ultimate effect of the substances is not detrimental to the health and welfare of the animal;

—The animals used, at present, for farming purposes should be preserved in a way that makes it possible to again start breeding a variety that may not have been bred for several years, should this be judged desirable.

Ekesbo concludes, "Man [sic] has always had ethical rules, written or unwritten, for animal husbandry. In our time with rapid scientific achievements, international agreement on ethical rules are necessary for the protection of the animals, the farmers and the society." While Ekesbo's proposal for ethical rules regarding biotechnology in farm animals seems reasonable, the rules and criteria are still subject to different interpretations depending on one's viewpoint. Unregulated ethical guidelines would be preferable to written rules in animal agriculture and in research, except for the most basic research. Anti–animal cruelty statutes, humane slaughter regulations and animal use guidelines for research already in existence are sufficient.

CONCLUSION

Animal agriculture contributes to the quality of human life by providing high–quality, nutrient–dense foods. Farmers have a moral obligation to produce this food as efficiently as possible. This will provide the maximum amount of human food while minimizing the consumption of natural resources and effects on the environment. Biotechnology should be used like any other tool to help achieve this goal. As the world population approaches six billion people, these persons' basic interest in being fed certainly takes precedence over the peripheral interests of animals served by over–regulation of animal production.

REFERENCES

Ekesbo, I. 1991. Biotechnology for control of growth and product quality in meat production, implications and acceptability for animal safety, health and welfare. In *Biotechnology for Control of Growth and Product Quality in Meat Production: Implications and Acceptability*. Proceedings of an International Symposium, Washington, DC. December 1990.

Gorner, P. 1992. Science meets profit motive in biotech industry. *The Chicago Tribune*. March 1, 1992.

Harrison, R. 1964. *Animal machines: the new factory farming industry*. V. Stuart, London.

Kaye, H. L. 1992. Are we the sum of our genes? *The Wilson Quarterly*. Spring. p. 68.

Kauffman, R.G. and B.J.E. Rutgers. 1991. The ethics of meat production: What man's relationship should be with animals and why. In *The European Meat Industry in the 1990's*. F.J.M. Smulders, ed. Elsevier, Amsterdam, The Netherlands. pp. 247–265.

National Agricultural Research and Extension Users Advisory Board. 1990. *Appraisal of the proposed 1991 budget for food and agricultural services*.

Report to the President and Congress, United States Department of Agriculture.
Van de Veer, D. 1979. Interspecific justice. *Inquiry.* 22:55–79.
Witt, S.C. 1990. *Biotechnology, microbes and the environment.* Center for Science Information. San Francisco, CA.

Bernard E. Rollin
Philosophy
Colorado State University

The Creation of Transgenic Animal "Models" for Human Genetic Disease

Perhaps the greatest socio–ethical challenges associated with the development and use of transgenic animals in biomedical research are the problems associated with animal welfare. Whereas the issue of biosafety does, indeed, represent a major concern, the minimization of such risk is as much a prudential concern as an ethical one for investigators, as they themselves are put at risk by failure to provide adequate safeguards against the dangers of transgenic animal research. Animal welfare concerns, on the other hand, represent a far greater moral challenge, for concern about animal welfare often does not coincide with perceived self–interest and, indeed, can exact costs in terms of self–interest, in the form of money, time, extra personnel, delay in research, etc. In other words, many researchers have traditionally not equated concern for animal welfare with self–interest and are, thus, unlikely to do the right thing for reasons of self–interest. Somewhat mitigating this blanket statement is the relatively recent acknowledgement of the fact that failure to assure animal welfare can skew variables relevant to research and actually compromise research (Rollin, 1990), but nonetheless, the coincidence of the two is far from perfect. As we shall see, certain aspects of transgenic animal research do represent an area where welfare could be ignored without obviously jeopardizing the work in question. Thus moral concern must take up the slack left after prudential considerations are exhausted.

The emergence of a systematic social ethic whose purview extends to the treatment of laboratory animals is a relatively recent phenomenon, as evidenced by the fact that researchers basically enjoyed *carte blanche* in the use of animals until the mid 1980s (Rollin, 1991). For most of the 19th and 20th centuries, the only consensus ethical principle extant in society for the treatment of animals was a prohibition against overt, willful, intentional, needless, wanton cruelty, as expressed in anti–cruelty legislation. Concerned as much with ferreting out sadistic individuals who might begin with animals and move to humans as with protecting animals, these laws, therefore, did not address "normal," "necessary" or "beneficial" sources of animal suffering such as agriculture, research, hunting, trapping or education; these are typically exempted from the anti–cruelty laws by statute, or else have been excluded by judicial decision. Rather, the laws focused on deviant behavior leading to "unnecessary" animal suffering. It is only in the past decade that society has begun to realize that a mere fraction of animal suffering is a result of overt cruelty—the vast majority of animal suffering at human hands, in fact, grows out of such decent motivations as increasing knowledge, curing disease, increasing efficiency of food production, protecting humans against toxic substances and so on. Correlative with this realization has come a demand for the control of suffering in areas of animal use which previously enjoyed *laissez faire,* notably toxicity testing, animal research and animal agriculture. First to be directly affected by this demand was animal research, with two major pieces of federal legislation designed to assure the welfare of research animals passed in the U.S. in 1985.

Perhaps the main feature of this legislation, which I have discussed at length elsewhere (Rollin, 1989; 1991), is a mandate to control pain, suffering and distress in research animals except where scientifically necessary, as in the study of pain, and even there, to minimize it as far as possible. Second, the legislation is designed to assure "enforced self–regulation" of animal research and dialogue about animal welfare concerns, through the vehicle of protocol and facilities review by animal care committees. Third, the legislation suggests that welfare concerns are not limited to controlling overt pain and suffering, but actually points towards providing some positive opportunity for animals to express their biological and behavioral natures—this is exemplified by the requirements of exercise for dogs and provision of an environment conducive to the psychological well–being of primates. This legislation has already had many salubrious effects on the welfare of laboratory animals, perhaps the most dramatic being the focusing of scientific attention on recognizing, characterizing and alleviating animal pain. It has also led researchers to far greater awareness of ethical questions in research, something which was traditionally stifled by widespread belief that science is and ought to be, "value–free" (Rollin, 1989).

Thus, we see the emergence of a new ethic for animals demanding, in essence, maximization of the interests of animals while they are being used for human benefit. The most articulate expression of this ethic thus far, has been the demand for the control of animal pain and suffering in research.

For certain aspects of transgenic animal use, this demand will be relatively easy to satisfy. Consider, for example, the patented Harvard mouse which is disposed to the development of tumors. In the words of the patent, this is "an animal whose germ cells and somatic cells contain an activated oncogene sequence introduced into the animal...which increases the probability of the development of neoplasms (particularly malignant tumors) in the animal" (U.S. Patent Number 4,873,191). Minimizing pain and suffering for such an animal is, in principle and in fact, no different from minimizing pain and suffering in nontransgenic animals in whom tumors are induced by other means: the establishment of endpoints for euthanasia, in terms of tumor size, so that the animal does not suffer, and the judicious use of anesthetics, analgesics and tranquilizers during operative or other procedures.

Similarly, there is no reason the second major thrust of the new social ethic cannot be applied to these transgenic animals—namely the provision of enriched environments and husbandry systems for these animals which allow them to actualize their behavioral and biological natures. In the case of transgenic mice, for instance, one should look to the recommendations outlined in literature on care of mice; for example, a recent article described a caging system for rodents that is meant to accommodate their behavioral needs (Sharmann, 1991). Indeed, the characterization of such environments and systems for a variety of animals is a primary purpose of the chapters in a book I am currently editing (Rollin and Kesel, in press). Thus, the vast majority of transgenic animals developed so far raise no additional welfare issues beyond those concerning nontransgenic laboratory animals.

Indeed, those welfare issues which are raised dramatically by transgenic animals are also continuous with analogous nontransgenic cases. I am referring to the creation and maintenance of seriously defective animals which are developed and propagated to model some human disease. This was traditionally accomplished through identification of adventitious mutations and selective breeding. Transgenic technology allows for accomplishing the same goal far more quickly and in a far wider range of areas. One can essentially replicate, in principle, any human genetic disease in animals—and therein lies the major ethical concern growing out of transgenic technology.

A recent chapter in a book devoted to transgenic animals helps to focus the concern:

> There are over 3,000 known genetic diseases. The medical costs as well as the social and emotional costs of genetic disease are enormous. Monogenic diseases account for 10% of all admissions to pediatric hospitals in North America... and 8.5% of all pediatric deaths.... They affect 1% of all liveborn infants... and they cause 7% of stillbirths and neonatal deaths. ... Those survivors with genetic diseases frequently have significant physical, developmental, or social impairment.... At present, medical intervention provides complete relief in only about 12% of Mendelian single–gene diseases; in nearly half of all cases, attempts at therapy provide no help at all (Karson, 1991.)

This is the context in which one needs to think about the animal welfare issues growing out of the use of transgenic animals in biomedical research.

On one hand, it is clear that researchers will embrace the creation of animal models of human genetic disease as soon as it is technically feasible to do so. Such models, which introduce the defective human genetic machinery into the animal genome, appear to researchers to provide convenient, inexpensive and—most importantly—high fidelity models for the study of the gruesome panoply of human genetic diseases outlined in the over three thousand pages of text comprising the sixth edition of the standard work on genetic disease, *The Metabolic Basis of Inherited Disease* (Scriver et al., 1989). Such "high fidelity models" may well reduce the numbers of animals used in research, a major consideration for animal welfare, but are more likely to increase the numbers as more researchers engage in hitherto impossible animal research. On the other hand, the creation of such animals can generate inestimable amounts of pain and suffering for these animals since genetic diseases, as mentioned above, often involve symptoms of great severity. The obvious question then becomes the following: Given that such animals will surely be developed wherever possible for the full range of human genetic disease, how can one assure that vast numbers of these animals do not live lives of constant pain and distress? Such a concern is directly in keeping with the emerging social ethic for the treatment of animals; as we said, one can plausibly argue that minimizing pain and distress is the core of recent federal legislation concerning animal use in research.

The very first attempt to produce an animal "model" for human genetic disease by transgenic means, as mentioned earlier, was the development, by embryonic stem cell technology, of a mouse which was designed to replicate Lesch–Nyhan's disease, or hypoxanthine–guanine–phosphororibosyl transferase (HRPT) deficiency (Hooper et al., 1987; Keuhn et al., 1987). Lesch–Nyhan's disease is a particularly horrible genetic disease, leading to a "devastating and untreatable neurologic and behavioral disorder" (Kelley and Wyngaarden, 1983). Patients rarely live beyond their third decade and suffer from spasticity, mental retardation and choreoathetosis. The most unforgettable and striking aspect of the disease, however, is an irresistible compulsion to self–mutilate, usually manifesting itself as biting fingers and lips. The following clinical description conveys the terrible nature of the disease:

> The most striking neurologic feature of the Lesch–Nyhan syndrome is compulsive self–destructive behavior—between 2 and 16 years of age, affected children begin to bite their fingers, lips and buccal mucosa. This compulsion for self–mutilation becomes so extreme that it may be necessary to keep the elbows in extension with splints, or to wrap the hands with gauze or restrain them in some other manner. In several patients, mutilation of lips could only be controlled by extraction of teeth.
>
> The compulsive urge to inflict painful wounds appears to grip the patient irresistibly. Often he [sic] will be content

> until one begins to remove an arm splint. At this point a communicative patient will plead that the restraints be left alone. If one continues in freeing the arm, the patient will become extremely agitated and upset. When completely unrestrained, he will begin to put the fingers into his mouth. An older patient will plead for help and if one then takes hold of the arm that has previously been freed, the patient will show obvious relief. If help is not forthcoming, a painful and often severe injury may be inflicted. The apparent urge to bite fingers is often not symmetrical. In many patients it is possible to leave one arm unrestrained without concern, even though freeing the other would result in an immediate attempt at self–mutilation.
>
> These patients also attempt to injure themselves in other ways, by hitting their heads against inanimate objects or by placing their extremities in dangerous places, such as in between the spokes of a wheelchair. If the hands are unrestrained, their mutilation becomes the patient's main concern and effort to inflict injury in some other manner seems to be sublimated (Kelley and Wyngaarden, 1983).

At the present, "there is no effective therapy for the neurologic complications of the Lesch–Nyhan's syndrome" (Stout and Caskey, 1988). Thus Kelley and Wyngaarden, in their chapter on HRPT–deficiency diseases, boldly suggest that "the preferred form of therapy for complete HRPT–deficiency (Lesch–Nyhan's syndrome) at the present time is prevention," i.e. "therapeutic abortion" (Kelley and Wyngaarden, 1983). This disease is so dramatic that I predicted almost a decade ago that it would probably be the first disease for which genetic researchers would attempt to create a model by genetic engineering.

Researchers have sought animal models for this syndrome for decades and have created rats and monkeys that will self–mutilate by administration of caffeine and other drugs (Boyd et al., 1965). Thus, it is not surprising that the first disease genetically engineered by embryonic stem cell technology was, indeed, Lesch–Nyhan's disease (Hooper et al., 1987; Keuhn et al., 1987). However, these animals were phenotypically normal and displayed none of the metabolic or neurologic symptoms characteristic of the disease in humans. The reasons for this are unknown (Stout and Caskey, 1988).

This case provides us with an interesting context for our animal welfare discussion. Although the animals were, in fact, asymptomatic, presumably at some point in the future researchers will be able to generate a symptomatic model transgenically. Let us at least assume that this can occur—if it cannot, there is no animal welfare issue to concern us! Whether one ought to create such animals is a question I have addressed elsewhere (Rollin, 1986). The practical moral question that arises is clear: Given that researchers will certainly generate such animals as quickly as they are able to do so, how can one assure that the animals live lives that are not characterized by the same pain and distress which they are created to model?

Again, this question does not differ in kind from the moral questions associated with developing traditional chronic animal models of human disease, be it by breeding, pharmacological manipulation or tissue destruction. The difference is in degree—transgenics provides the potential for generating vast numbers of animals modeling genetic diseases with devastating symptoms. A second difference lies in the fact that transgenic technology is developing at precisely the same time that social/ethical demand for controlling pain and suffering in research animals is at its historical peak and seems to be increasing.

Regrettably, researchers in the past have been cavalier in controlling pain and suffering in animals used as chronic disease models. Though many of the animals have required extraordinary amounts of care and husbandry, such efforts have been directed, for the most part, at keeping the animals alive and scientifically functional rather than at controlling pain and suffering. Given our current social ethic, it is increasingly imperative that pain and suffering be controlled in all animals used for research. Thus, concern for this dimension of animal care needs to be a fundamental principle which guides those contemplating the transgenic creation of animals which replicate human genetic disease. Such an issue is a true moral challenge for researchers, as concern for the animals' quality of life will undoubtedly make things more difficult and expensive for researchers. At the same time, it is patent that such concern is both morally and socially obligatory. Furthermore, failure to assure the public that animal suffering is being minimized could well accelerate major political constraints on all areas of biotechnology (Rollin, 1986).

Unfortunately, because the research community traditionally ignored this moral component of animal research, there is no vast literature on controlling pain and suffering in chronically defective animals. There has probably been more scientific attention to such questions during the six years following the passage of the aforementioned federal legislation than in the entire previous history of animal research (Rollin, 1989). Doubtless such attention will continue to grow at a significant rate. Researchers undertaking work with animals which model human genetic disease should, therefore, vector these concerns into protocol planning and budgeting; funding agencies should demand such planning, and animal care and use committees should not approve projects until they have evidenced that pain, suffering and distress are controlled.

In many cases—perhaps in a symptomatic Lesch–Nyhan's animal—management of suffering may require a far more radical approach than the standard uses of anesthesia, analgesia and tranquilization, which are, by and large, used for short periods of time. If a defective animal is to be kept alive for long periods and is likely to experience pain and suffering during that period, researchers should consider the possibility of effecting total elimination of consciousness. One such approach could involve surgically rendering an animal decerebrate, so that, while vegetative functions are extant, the animal's subjective experience has been shut down. Alternatively, and perhaps more viably, one could render an animal irreversibly comatose so that it was effectively anesthetized throughout its life. Unfortunately, virtually no literature exists on induction of coma.

I have galvanized a team of researchers at Colorado State University to explore this drastic possibility. We utilize animals scheduled to be euthanized for other reasons and attempt to induce irreversible coma in these animals by induction of cerebral hypoxia. We hope to find a clear EEG criterion which signals coma. If the method is successful, perhaps the method could be taught to veterinarians at institutions planning to utilize animal models of genetic disease so that the animals will not needlessly suffer.

Obviously, such methods of controlling pain and suffering are very drastic and their effective application is fraught with difficulties. For example, they could, presumably, only be employed where higher brain function is essentially irrelevant to the study of the disease. Whether this is the case or not with Lesch–Nyhan's disease, for example, once it was established that the transgenic animal, indeed, showed all signs of the disease, is unclear. I believe it is. Certainly, at least some metabolic genetic diseases could be studied in this way.

Equally significant, there is something aesthetically, at least and perhaps morally as well (I am not clear on this), about deliberately creating such animals. At the very least, it dramatically perpetuates the notion that society is seeking to transcend—that animals are simply tools for human expedient use. It is, in my view, the lesser of two evils.

The key point is that this dimension of genetic engineering of animals cannot be ignored. There is, as we saw, every reason to believe that transgenic animals will be created to study human genetic disease as soon as the technological capability exists to do so. Extant laws permit such animals to be created. The mindset of the research community makes it inevitable. It is also clear that such diseases can cause enormous amounts of pain and suffering. In the face of this development, responsible researchers need to explore all possible avenues for controlling such pain and suffering. These approaches should include such established methods as the liberal use of anesthetics, analgesics and tranquilizers, and by making as much of the research as possible acute. But these methods are unlikely to be effective in the case of those diseases where suffering begins at birth or is chronic after a certain stage of development. (Lesch–Nyhan's patients, as we mentioned, do not show symptoms from birth, but do exhibit them chronically after their onset.) Thus, methodologies need to be developed which will control pain and suffering over extended periods of time. There is, thus far, no reason to believe that the research community has yet engaged this issue *vis à vis* animals used in other chronically painful work, let alone in genetically engineered animals. The development of such methodologies for controlling pain and suffering is likely to be exportable to numerous areas of animal research, not only transgenic creation of disease. Only in this way can research attempt to stay in harmony with the ethical stance of the society which allows and supports it.

REFERENCES

Boyd, E.M., M. Dolman, L.M. Knight and E.P. Sheppard. 1965. The Chronic Oral Toxicity of Caffeine. *Can. J. Physiol. and Pharm.* 43:95.

Hooper, M., K. Hardy, A. Handyside, S. Hunter and M. Monk. 1987. HPRT–Deficient (Lesch–Nyhan) Mouse Embryos Derived from Germline Colonization by Cultured Cells. *Nature.* 326:292.

Karson, E.M. 1991. Principles of Gene Transfer and the Treatment of Disease. In *Transgenic Animals.* N. First and F.P. Haseltine, eds. Butterworth–Heinemann, Boston.

Kelley, W.N. and J.B. Wyngaarden. 1983. Clinical Syndromes Associated with Hypoxanthine–Guanine Phosphororibosyltransferase Deficiency. In *The Metabolic Basis of Inherited Disease, fifth edition.* J.B. Stanbury, J.B. Wyngaarden, D.S. Fredrickson, J.L. Goldstein, and M.S. Brown, eds. McGraw–Hill, New York.

Kuehn, M.R., A. Bradley, E.J. Robertson and M.J. Evans. 1987. A Potential Model for Lesch–Nyhan Syndrome Through Introduction of HPRT Mutations into Mice. *Nature.* 326:295.

Rollin, B.E. 1986. The Frankenstein Thing. In *Genetic Engineering of Animals.* J.W. Evans and A. Hollaender, eds. Plenium Press, New York.

Rollin, B.E. 1989. *The Unheeded Cry: Animal Consciousness, Animal Pain and Science.* Oxford University Press, Oxford.

Rollin, B.E. 1990. Ethics and Research Animals–Theory and Practice. In *The Experimental Animal in Biomedical Research, Volume I.* B.E. Rollin and M.L. Kesel, eds. CRC Press, Boca Raton, FL.

Rollin, B.E. 1991. Federal Laws and Policies Governing Animal Research: Their History, Nature and Adequacy. In *Biomedical Ethics Reviews 1990.* J.M. Humber and R.F. Almeder, eds. Human Press, Clifton, N.J.

Rollin, B.E. and M.L. Kesel, eds. in press. *The Experimental Animal in Biomedical Research, Volume II.* CRC Press, Boca Raton, FL.

Scriver, C.R., A. L. Beaudet, W. S. Sly and D. Vale, eds. 1989. *The Metabolic Basis of Inherited Disease, Volumes I and II.* McGraw Hill, New York.

Sharmann, W. 1991. Improved Housing of Mice, Rats and Guinea Pigs: A Contribution to the Refinement of Animal Experiments. *Atla.* 19:108.

Stout, J.T. and C.T Caskey. 1988. Hypoxanthine Phosphororibosyltransferase Deficiency: The Lesch–Nyhan Syndrome and Gouty Arthritis. In *The Metabolic Basis of Inherited Disease, Volume one.* C.R. Scriver, A.L. Beaudet, W.S. Sly and D. Vale, eds. McGraw–Hill, New York.
United States Patent number 4,873,191, October 10, 1989.

Dale Jamieson
Philosophy
University of Colorado
George E. Seidel Jr.
Animal Science
Colorado State University

Workshop Report

The workshop began with a wide–ranging discussion about the concept of well–being. To what extent does it overlap with productivity? Does it take into account the sentience of the animal, the exercise of its cognitive capacities and its ability to cope with its environment? It was asserted that at the least, well–being involves an ability to express a range of normal behaviors. It was remarked that well–being is not a single, precisely defined state. The indicators of well–being are not linear. In some cases, more of something may be better, but in other cases there may be thresholds such as optimal body weight. Perhaps well–being should be defined negatively rather than positively: the failure of some system may imply a lack of well–being, but the presence of something (e.g., normal blood glucose levels) may not indicate an acceptable level of well–being. Productivity (e.g., growth, milk production, reproduction) has been used as one measure of well–being for agricultural animals although it was noted that such measures often will be inappropriate for determining the well–being of pets and research animals. It was also remarked that the notion of well–being applies to individuals, but indices of well–being are relative to populations.

Participants, faced with a vast scope of issues in animal well–being, decided that the discussions should be limited to the well–being of animals involved in biotechnology: farm animals and experimental animals. What are the factors that we should look for to measure well–being in transgenic animals? What about the integrity or intrinsic value of the animal? It was suggested that there were two problems in proceeding this way. First, not all concerns about the use of animals involve inadequate well–being during these uses. Rather, some individuals object to specific uses of animals. Perhaps harms that animals may suffer as a result of biotechnology should be considered. Discussion could focus on whether infliction of these harms can be justified, or whether there are alternative procedures. Secondly, it was pointed out that biotechnology involves much more than just transgenic procedures. Selective breeding, artificial insemination and embryo transplantation are other examples. Different forms of biotechnology may raise different issues. Animals may be harmed deliberately in order to create disease models for research purposes, whereas any harms animals suffer in production are incidental and usually not intended.

These points led to a discussion of why older forms of biotechnology did not stimulate as much public concern as newer forms. Selective breeding has been going on for centuries without serious well–being objections. Many people feel that new technologies create new problems. Furthermore, societal views of animals are changing. Technology is colliding with changing morality. In addition, lay people are increasingly skeptical of scientific ethics, due in part to recent negative publicity associated with a few famous scientists.

Some people are upset about biotechnology because it is perceived as "unnatural." This connects to general cultural attitudes about the human relationship to nature. Against this it was suggested that people are part of nature, so everything we do is natural. The objection to this, however, is that this concept eludes the distinction between the natural and unnatural.

It was suggested that one new thing about contemporary biotechnology is that we can create animals who may have no potential for "happy" lives. Can we be said to have wronged such animals?

Some participants suggested that biotechnology gives us the capability to create animals that are adapted to a closed–confinement system in which they cannot move or otherwise express some movement–related behaviors. We may be able to imagine creating animals that would delight in confinement—to some extent this has been done already. Some thought that it would be wrong to do this because it would be exploiting the animals, although it was remarked that this objection was not an objection on grounds of animal well–being.

QUESTIONS RAISED ABOUT BIOTECHNOLOGY AND ANIMAL WELL–BEING

Individual participants listed questions about biotechnology and animal well–being during the workshop. In no specific order they are:

—Will the farm community compensate for bad environments by creating new animals?

—Will genetic engineering arbitrarily be singled out from other forms of genetic manipulation (e.g., pet breeding, spaying, castration)?

—Will biotechnology be used to reduce and replace animals used in painful and invasive research?

—Are we changing our attitude and relationship towards nature, e.g., from the standpoint of mastering nature?

—Will a concern with animal well–being lead us to overlook the important human issues involved in the use of biotechnology?

—Whose interests are being served by biotechnology?

—Does biotechnology raise new issues about animal well–being? How much has the technology changed and how much have we changed?

—Are societal needs or scientific needs driving the development of biotechnology?
—What do we want the future animal production system to be like and how can biotechnology be used to forward our vision?
—Why is it difficult to pose fundamental questions about biotechnology?
—Why do people respond so strongly to biotechnology?
—Are there nonbiotechnological alternatives to what we want to do?
—Is biotechnology different in kind from what we have been doing?
—How does money flavor the issue?
—What are the larger beliefs about the relations between humans and animals that drive the differences in opinion about biotechnology? (It was observed that the emergence of biotechnology coincides with greater concerns about animals, increasingly cognitive views of animals, increased distance from agricultural and draft uses of animals, and urbanization and romanticization of animals. Genetic engineering feeds into these concerns because of the general concern that the manipulation of genes could lead to unnatural beings).

CONCERNS ABOUT BIOTECHNOLOGY AND ANIMAL WELL-BEING

Individual workshop participants had concerns which were then listed as follows (again in no special order):

—Some of the public believe that there are more costs than benefits to biotechnology;
—Biotechnology could increasingly depersonalize the relationship between people and animals;
—Everything we do to animals biotechnologically could eventually be done to humans (and thought to be acceptable because of the previous animal work);
—Politicians frequently do not understand the social ramifications of their actions;
—Scientists and farmers want to make decisions about the use of biotechnology (even though they do not always understand the implications of their decisions) and some resent the intrusion of lay people into routine decision–making;
—People like us (well–educated, predominantly urban, well–placed) are making certain decisions for rural farmers and there is little they can do about it. Biotechnology and its regulation are being driven by intellectual power rather than by people who work with animals on a daily basis;
—Uncertain, and by some interpretations frivolous, restrictions may be imposed that will prevent society from reaping the benefits of biotechnology;
—Uncertain, and by some interpretations frivolous, use of biotechnology may limit the options of society to have a just and sustainable future;
—There is a risk that biotechnologists will defend the use of animals so avidly that they will evoke a huge counter–reaction;
—Some critics of biotechnology want to stop all human uses of animals;

—The rate of change goes beyond our ability to anticipate it and to respond to it;
—People do not understand that animal biotechnology could eventually lead to the elimination of animal food production (through fermentation technologies, etc.);
—There is a lack of effective communication between science and the public;
—Genetic engineering may eventually make the discussion of animal well–being a nonissue;
—The failure to use animals in genetic engineering will deprive people of the benefits of medical pharmaceuticals;
—If we cannot predict the impacts of genetic engineering on animal well–being, should we still undertake it?

By this point in the discussion the participants were quite polarized and it was remarked that some people who hold very strong views were not even represented. It was suggested that some of the fundamental questions might concern whether it is ever acceptable to utilize animals for human use, whether animal biotechnology poses unique questions about animal well–being and whether biotechnology is qualitatively or quantitatively different from what has come before.

POSSIBLE HARMS OF BIOTECHNOLOGY TO ANIMAL WELL–BEING

Participants decided to identify some possible harms and benefits to animal well–being that arise in the context of biotechnology. Possible harms elicited from participants were:

—Loss of genetic diversity;
—Proliferation of genetically defective animals who suffer as disease models;
—Thinking of domestic animals as human artifacts;
—Diverting resources away from improving traditional husbandry practices;
—Leading us away from sustainable agriculture which might be better for animals;
—Leading to ecological devastation through the introduction of genetically altered animals;
—Narrowing the concept of well–being to merely biological health;
—Strengthening corporate agribusiness with long–term negative consequences for animal well–being;
—Creating animals who do not feel pain and may still damage themselves.

POSSIBLE BENEFITS OF BIOTECHNOLOGY TO ANIMAL WELL–BEING

Following a listing of harms to animal well–being, workshop participants then listed possible benefits biotechnology could bring to animal well–being:

—Removal of genetic defects from animal populations more rapidly;
—Permitting increased disease resistance;
—More efficient production leading to the use of fewer animals;

—Better understanding of animal well–being;
—Better well–being through the creation of animals less susceptible to environmental conditions;
—Creating animals so existing resources can be used as food;
—Using fewer animals both in research and in farming;
—Increasing genetic diversity;
—Increasing understanding of and solutions to the medical problems of both humans and nonhumans;
—Enhancement of wildlife management and growth of nonhuman populations as hunting becomes obsolete;
—The end of factory farming through the redesign of farm animals;
—Driving small producers out of business who mistreat animals.

A highly unrepresentative straw poll was then done in order to see which of these possible harms and benefits the participants most wanted to discuss. As listed, the first four possible harms and the first four possible benefits received the most support along with the possibility that animal biotechnology may lead to healthier products for both humans and animals. They formed the basis of much of the remaining discussions.

CONSENSUS STATEMENTS

Weighing the broad spectrum of issues related to biotechnology and animal well–being, participants were able to reach agreement on four consensus statements:

1. Biotechnology may contribute to animal well–being, but it is not the only approach to improving animal well–being.

2. There should be responsible, systematic investigation of the benefits and harms to animals that may be associated with biotechnology.

3. It is acceptable under some conditions to use animals for human use.

4. Animal biotechnology has the potential to contribute to the "three Rs" in animal experimentation: reduction, refinement and replacement.

RECOMMENDATIONS

1. With respect to animal well–being, criteria should be developed for responsible research and application of specific biotechnologies in animals. The full spectrum of opinion should be represented in the development of these criteria. These criteria should be periodically reconsidered in the light of changing circumstances.

2. The benefits and harms noted should be taken into account in developing these criteria.

3. Animal biotechnology should not be used in ways that impose great costs in animal well–being, while achieving only minor human or animal benefits. When there is the likelihood that a procedure will cause great suffering to animals, alternatives should be sought.

LINKS OF ANIMAL BIOTECHNOLOGY TO HUMAN HEALTH

Clifton A. Baile
Distinguished Fellow and Director
Research and Development
Monsanto Company

The Potential Spinoff of Advances in Human Medicine to Animal Research and Agriculture

BIOTECHNOLOGY AS A GROWTH INDUSTRY

The discoveries of biotechnology will soon be approaching 20 years of age. Several of the key discoveries—restriction enzymes (Kelly and Smith, 1970), DNA ligases (Weiss et al., 1968), etc. led to the first transformed microbe in 1973 (Cohen et al., 1973). These initial discoveries led to an extended discussion, in 1976, in a San Francisco tavern between a venture capitalist Robert A. Swanson and Herbert W. Boyer of the University of California in San Francisco. The first biotechnology–based "boutique," Genentech, was thus soon formed. Genentech's success led the way to the creation of hundreds of other small companies specializing in applications of an emerging technology—genetic engineering. The excitement, dreams and speculation generated by this technology attracted billions of dollars of venture capital, primarily for human medical interests. By January 1, 1992, Amgen, one company of these origins, joined the Standard & Poor's 500 and was given a $10 billion market evaluation.

The growing market of biotechnology–based protein drugs is now well–over $2 billion annually. In the late 1970s, interest in animal applications of biotechnology started to attract investments. Bovine somatotropin (BST) product development was the first direct spinoff of this early drug research activity. Many other applications of biotechnology for animals and plants have subsequently been the basis of the formation of small companies. Federal agencies directed close to $4 billion to support biotechnology research in 1992; this is of special interest to those of us interested in animal production and the development of improved methods for enhancing the efficiency of production. Even more than that is being spent by hundreds of companies to develop further scientific bases and applications of biotechnology. Of tremenendous potential for broadening the horizons for many aspects of biological research is the 15–year, $3 billion federal support of the Human Genome Project.

BIOTECHNOLOGY ADVANCES IN HUMAN MEDICINE AND SPINOFFS TO ANIMAL AGRICULTURE

Others in this conference will undoubtedly discuss the numerous and essential contributions of animal models to the development of human drugs and treatment strategies for even the most intractable of the human disease conditions. The following discussion will deal with certain aspects of the applications of biotechnology to animal research and agriculture.

In Table 1 is a list of protein drugs approved by the Food and Drug Administration (FDA)since 1982, starting with Humulin®, a biosynthetic insulin (*Genetic Engineering News*, January, 1992). Momentum in protein drugs is growing—5 of the 30 products approved by the FDA in 1991 were of this type. This rate may be duplicated in 1992 and possibly 100 more biotechnol-

TABLE 1: BIOTECHNOLOGY-BASED DRUGS APPROVED BY FDA

Product	Company	Indication	Year
Actimmune	*Genentech*	*management of chronic granulomatous disease*	*1990*
Activase	*Genentech*	*acute myocardial infarction*	*1987*
		acute pulmonary embolism	*1990*
Alferon	*Interferon Sciences*	*genital warts*	*1989*
Engenix	*SmithKlineBeecham*	*hepatitis B*	*1989*
Epogen	*Amgen 2*	*treatment of anemia associated with chronic renal failure, including patients on dialysis & not on dialysis, and anemia in Retrovir-trated HIV-infected patients*	*1989*
Procrit	*Ortho Biotech*	*treatment of anemia associated with chronic renal failure, including patients on dialysis & not on dialysis, and anemia in Retrovir-trated HIV-infected patients*	*1990*
Humatrope	*Eli Lilly*	*human growth hormone deficiency in children*	*1987*
Humulin	*Eli Lilly*	*diabetes*	*1982*
Intron	*Schering-Plough*	*hairy cell leukemia*	*1986*
		genital warts	*1988*
		AIDS-related Kaposi's sarcoma	*1988*
		non-A, non-B hepatitis	*1991*
Leukine	*Immunex*	*autologous bone marrow tranplantation*	*1991*
Prokine	*Hoechst-Rooussel*	*autologous bone marrow transplantation*	*1991*
Neupogen	*Amgen*	*chemotherapy-induced neutropenia*	*1991*
Orthoclone OKT3	*Ortho Biotech*	*reversal of acute kidney transplant rejection*	*1986*
Protropin	*Genentech*	*human growth hormone deficiency in children*	*1985*
Recombovax HB	*Merck*	*hepatitis B prevention*	*1986*
Roferon-A	*Hoffman-LaRoche*	*hairy cell leukemia*	*1986*
		AIDS-related Kaposi's sarcoma	*1988*

ogy–based drugs are now in clinical trials. Applications of identical or similar products for animals are bound to follow all of this activity. A major advantage of this is that human drug development leads to initial reagents and test probes for studies both *in vitro* and *in vivo* in other animals.

Of longer–term consequence is the development of a large number of products resulting in significant production process developments and associated discoveries which lead to reduced costs of drug production. In many cases, this alone permits consideration of new animal product concepts. Additionally, some of these biosynthetic proteins are likely to have desirable effects during certain physiological states of animals not predicted from their names or the initial basis for their discovery.

The rapid development of genetic engineering has made it possible for a soaring rate of new discoveries. Proteins present in minute quantities, but occasionally of immense importance to animal physiology, can be predicted from messenger RNA. Thus, the existence of previously unknown proteins can be demonstrated by isolation and multiplication of very specific genetic codes. This has led to the discovery of numerous important proteins (i.e., hormones, receptors, enzymes) that otherwise would have been impossible to discover by classical endocrinology–based techniques, even before knowing the function of these proteins.

Within the lifetimes of many of us, the understanding of receptors on cells have evolved from essentially a concept to a specific protein or a family of proteins. These advances have, in many cases, replaced the need for pharmacological classifications for receptor types such as the adrenergic receptors which are subdivided according to their pharmacological response to epinephrine and non–epinephrine (Ahlquist, 1948). The genetic expression of a hormone receptor and binding characteristics of each of a family of receptor proteins can now be studied for cell type specificity, etc. (Laird et al., 1991). The diversity of protein receptors has led to new strategies for drug design including the utilization of very powerful chemical tertiary structure software for predicting specific drug analogs with highly selective activity. In some cases, possession of quantities of a specific receptor protein allows screening for specific binding by peptides or other chemicals to develop specific blockers.

As recently reported (Gibbons, 1992), second generation products from biotechnology will include other specific means of modifying selected protein expression. The expression of specific proteins which either inhibit growth enhancement or cause disease conditions may be reduced or eliminated by blocking the transcription of DNA into the specific protein messenger RNA. Another method of obtaining similar responses is to develop short oligonucleotides that recognize and bind to specific messenger RNAs and thus block protein synthesis. These and other means provide opportunities for developing drugs that enhance animal production and health.

SUMMARY

Human health–related research is leading the way to new technological breakthroughs, reagents, probes and general understanding of biological systems at a molecular level. An additional benefit will be the generation of many opportunities for enhancing animal production. In many cases, species specificities may be engineered into the developing strategies to ensure that human health concerns are met even at a molecular level. At the present time there are more opportunities for improving the efficiency and endproduct quality (e.g., nutritional value), than can be funded by most major companies. The many exciting advances in biotechnology are steadily making the discovery process more affordable and allowing the backlog of opportunities to build. Recognition and acceptance of how to safely apply biotechnolgically based productivity enhancers and health aids will surely result in much greater investment in animal drug development in the near future. All of this activity, plus the advances in animal science, will assure the world new methods of increasing the abundance, quality and variety of foods. Given the prediction that 10 billion people may be on earth within the first half of the 21st century, this is an issue of great importance. These future products will be cheaper and safer to produce and will lead to superior, safer and environmentally friendly management options for animal agriculture. All of these qualities should make these innovations available to more of the world's food producers including those in developing countries.

REFERENCES

Ahlquist, R.P. 1948. A study of the adrenotropic receptors. *Am. J. Physiol.* 153:586.

Biotechnology–based Drugs Approved by FDA (Table). 1992. *Genetic Engineering News*, January.

Cohen, S.N., A.C.Y. Chang, H.W. Boyer and R.B. Helling. 1973. Construction of biologically functional bacterial plasmids in vitro. *Proc.Natl. Acad. Sci. USA.* 70:3240.

Kelly, T.J. and H.O. Smith. 1970. A restriction enzyme from hemophilus infuenzae II: base sequence of the recognition site. *J. Mol. Bio.* 51:393.

Gibbons, A. 1992. Biotech's second generation. *Science.* 256:766.

Laird, D.M., D.P. Creely, S.O. Hauser and G.G. Krivi. 1991. Analysis of the ligand binding characteristics of chimeric human growth hormone/bovine growth hormone binding proteins. Endocrin. Soc. 73rd Annual Meeting, (Abstr 1545).

Weiss, B., A. Jacquemin–Sablon, T.R. Live, G.C. Fareed and C.C. Richardson. 1968. Enzymatic breakage and joining of deoxyribonucleic acid. *J. Biol. Chem.* 243:4543.

Fuller W. Bazer
Animal Science
Texas A&M University

Reproductive Biology of Domestic Animals: Linkages with Veterinary and Human Medicine

This paper will discuss opportunities for linkages between research in animal agriculture and research directly related to human medicine. Results of research in animal agriculture have affected certain aspects of clinical medicine for many years. The application of biotechnology to human and veterinary medicine involves many techniques used in reproductive biology and applied in animal agriculture. Among these are embryo transfer, gene injection, use of embryonic stem cells for introduction of genes and cloning of embryos. Gene transfer in animals is being used to evaluate the value of transgenic animals in animal agriculture. Transgenic animals are also being used for the study of specific genetic defects. Production of transgenic animals has received much attention recently. This technique represents only one approach which can be used not only to examine questions of interest to the biomedical community, but will also help establish linkages between animal agriculture, veterinary medicine and human medicine. Naturally occurring diseases in livestock must be examined for their value as models for human disease. Regulatory proteins normally secreted by the conceptus (embryo/fetus and associated membranes) may have unique therapeutic value for certain diseases. In addition, regulatory proteins secreted in significant amounts may be associated with normal development of the conceptus, but with various disease states in adults. Although it may be useful to create transgenic animals as models for specific diseases, we must also focus on naturally occurring diseases affecting animals that are common to humans.

Reproductive biology, in particular the study of pregnancy in livestock, provides numerous opportunities to address questions of biology with application to both veterinary and human medicine. Pregnancy and associated development of the conceptus seems analogous to compressing events of a lifetime into a period of gestation. Because of extremely rapid development of the conceptus, basic questions relative to proliferation and differentiation of cells, cell–cell interactions and regulation of gene expression can be addressed. Results impact our knowledge of wound healing, cancer biology, tissue transplantation biology, immunology, developmentally regulated expression of genes, endocrine regulation of maternal and fetal–placental expression of genes, the hematopietic system and so forth. It is not surprising that investigators from numerous disciplines have petitioned for the opportunity to study fetal–placental tissues of humans. Again, understanding normal events and mechanisms associated with the reproductive process will directly impact animal agriculture and the biomedical community.

GENE TRANSFER IN HUMAN AND NONHUMAN ANIMALS

It is clear that techniques such as embryo transfer, gene injection, cloning of embryos and related technologies have always been done first in animals so that proven methods can later be applied to human and veterinary medicine. Gene transfer in animals is now being evaluated critically in a number of species and with a number of genes. The problems laboratories face in attempting to produce transgenic animals include poor expression, lack of expression or over expression of genes of interest, and in some cases, lack of incorporation of injected genes. Documentation of these problems in animals and development of technologies to overcome them must be through animal experimentation before these techniques become useful to clinical medicine. In some cases, tissue specific expression is essential. This is also being studied extensively in animals in which expression of genes are being restricted to the mammary gland and the gene products are being harvested from milk. These technologies will eventually impact animal agriculture and medicine, but experimentation with animals will continue to be central to this research.

ALTERNATIVES TO TRANSGENIC ANIMALS

An alternative to development and use of transgenic animals is the identification of animal models which have naturally occurring diseases and/or metabolic disorders that provide natural linkages between animal agriculture and the biomedical community. Although rodents or other small laboratory animals are commonly used, domestic animals should be used as models whenever possible so that scientific breakthroughs will impact animal health, human health and production animal agriculture. For example, sheep are used extensively as the animal model for studies of basic questions pertaining to pregnancy, perinatology and neonatology in humans in departments of pediatrics and departments of obstetrics and gynecology. Results from such research directly impacts the base of knowledge available to clinicians to improve survival and well–being of neonatal humans and animals. Importantly, the same information benefits production animal agriculture.

Trophoblast Interferons

Acquired immunodeficiency syndrome (AIDS) is having devastating effects on human health throughout the world. Similar lentivirus–induced diseases affect cattle, sheep, horses, goats, primates and cats which makes them excellent animal models for studies of the etiology, prevention and treatment of human AIDS. A potential therapeutic drug for AIDS is recombinant leukocyte interferon alpha. The trophoblast of conceptuses of some species (e.g., ruminants and humans) produce unique interferons which may be especially useful for treatment of AIDS and AIDS–like diseases because the trophoblast interferons appear to lack the undesirable toxic side effects of leukocyte interferons.

Type I trophoblast interferons (tIFN) of sheep, goats and cows are biochemical signals for maternal recognition of pregnancy which may be useful for enhancing fertility in animal agriculture; however, they may also have therapeutic value in human and veterinary medicine (Bazer and Johnson, 1991). The tIFNs have high amino acid sequence homology with both interferon alpha I ($IFN_{\alpha I}$) and interferon alpha II ($IFN_{\alpha II}$) or omega interferon (IFN_{Ω}) which are produced by white blood cells. The tIFNs produced by sheep, cow and goat conceptuses are very similar to each other in structure and biological activity. A gene for human tIFN (htIFN) that has 85 to 87 percent homology with sheep tIFN (otIFN) has been cloned from a human placental cDNA library (Whaley et al., 1991).

The tIFNs have potential therapeutic value because they are interferons that inhibit cellular proliferation (Pontzer et al., 1991), exert antiviral effects (Pontzer et al., 1991) and regulate the immune system (Newton et al., 1989). Evaluation of the potential therapeutic value of tIFNs requires sufficient amounts of pure protein for clinical studies. A synthetic gene for sheep tIFN is being used in yeast and bacterial expression systems to produce recombinant sheep tIFN (rotIFN) that is identical to natural otIFN in terms of its amino acid sequence and biological activities (Ott et al., 1991). The antiviral activity of this tIFN is as potent as that of known recombinant leukocyte interferons from humans (rhIFN) and cattle (rbIFN), but sheep recombinant tIFN does not exert cytotoxic effects characteristic of treatment with $rhIFN_{\alpha}$ and $rbIFN_{\alpha}$ (Pontzer et al., 1991; Bazer et al., 1989). Exposure of human and feline peripheral lymphocytes infected with human immunodeficiency virus (HIV) and feline immunodeficiency virus (FIV) respectively, to sheep tIFN inhibited replication of the viruses, but did not exert cytotoxic effects on the infected cells when used at concentrations up to 200,000 antiviral units per ml. However, $rhIFN_{\alpha}$ and $rbIFN_{\alpha}$ exerted significant cytotoxic effects at only 1,000 to 5,000 antiviral units per ml.

The tIFNs also have antiproliferative effects on cells that is equivalent to or greater than that of $rbIFN_{\alpha}$ and $rhIFN_{\alpha}$ (Pontzer et al., 1991) and may be useful in the treatment of cancers. When anticellular activities of sheep tIFN, $rbIFN_{\alpha}$ and $rhIFN_{\alpha}$ were compared using human amnion (WISH) and Madin–Darby bovine kidney (MDBK) cells, all inhibited proliferation of the cells. However, sheep tIFN was more effective at lower dosages and, at high dosages (50,000 antiviral units per ml), otIFN more effectively blocked cell proliferation without adverse effects on cell viability. At the same concentrations of antiviral activity, rbIFN$_{\alpha}$ caused substantial cell death (Pontzer et al., 1991; Bazer et al., 1989).

Human patients having steroid–dependent mammary tumors respond to treatment with β–interferons because of increased receptors for progesterone and decreased receptors for estrogen in tumor cells (DeCicco et al., 1988). In the pregnant uterus of sheep, tIFN stabilizes receptors for progesterone while decreasing receptors for estrogen (Ott et al., 1992). The health of many women is affected adversely by estrogen–dependent tumors of the mammary glands and reproductive tract. Because estrogen–dependent tumor growth depends on the presence of cellular receptors for estrogen, tIFNs have potential therapeutic value because they inhibit synthesis of cellular receptors for estrogen which should prevent estrogen–dependent growth of the tumors.

Humans suffering from infection with HIV or diseases such as hairy cell leukemia are willing to consider lifelong therapy with rhIFN; however, chronic treatment with rhIFN results in development of resistance to the effects of currently available recombinant rhIFN (Tamm et al., 1987). In addition, high doses of rhIFN produce intolerable fever and chills, anorexia, weight loss and fatigue (Oldham, 1985) and may also cause seizures (Janssen et al., 1990). Interferons have both immuno–enhancing and cytotoxic effects; therefore, therapeutic doses are chosen which favor the immuno–enhancing effects. In contrast, tIFNs act through receptors on the uterine epithelium which are in direct contact with the conceptus and are exposed to as much as 40 million units of antiviral activity per 24 hours without cytotoxic effects. These tIFNs have unique "cell friendly" properties which may make them especially desirable therapeutic agents for use in animal agriculture, veterinary medicine and human medicine.

A number of diseases affecting livestock result from infections by lentiviruses of the family *Retroviridae*. These include, ovine progressive pneumonia virus (OPPV), caprine arthritis–encephalitis virus (CAEV), bovine immunodeficiency–like virus (BIV), equine infectious anemia (EIA), feline immunodeficiency virus (FIV) and simian immunodeficiency virus (SIV) (Haase, 1986). Diseases caused by OPPV, CAEV and BIV, for example, are uniquely suited for testing the therapeutic value of tIFNs in the control of lentivirus–induced diseases because conceptuses of each of these species secrete tIFN. These animal models must be studied to assess the therapeutic value of tIFNs in preventing or ameliorating vertical transmission of lentiviruses (i.e., via the placenta), and horizontal transmission (i.e., animal to animal), as well as the efficacy of tIFNs in treating infected fetuses and adult sheep, goats and cattle.

Hematopoietic System

Another protein isolated from reproductive tissues of domestic farm animals which has potential application to veterinary and human medicine is uteroferrin. Uteroferrin is a purple–colored, progesterone–induced glycoprotein secreted by uterine endometrial epithelium of pigs (Bazer et al., 1991). Uteroferrin can also be purified from human term–placenta. During pregnancy, uteroferrin is transported from uterine secretions into the fetal–placental circulation and is targeted to reticuloendothelial cells of the fetal liver, the major site of hematopoiesis in fetal pigs. Uteroferrin, from pig uterus is a tartrate–resistant acid phosphatase with many properties in common with the Type 5 tartrate–resistant acid phosphatase in human placenta, chondrocytes of osteoclastic bone tumors, spleens of patients with hairy cell leukemia, as well as Gaucher's and Hodgkin's diseases (Ketcham et al., 1985). In addition, uteroferrin has characteristics similar to those for purple acid phosphatases from normal bovine, rat, mouse and pig spleen, as well as bovine milk, bovine uterine secretions, equine uterine secretions and rat bone. Uteroferrin from pig uterus and human placenta is a hematopoietic growth factor having granulocyte–erythrocyte–monocyte/macrophage–megakaryocyte colony forming unit (CFU–GEMM) activity that affects differentiation of primitive nonadherent hematopoietic stem cells from pig and human bone marrow (Bazer et al., 1991).

Uteroferrin from pig uterine endometrium and human placenta (Ketcham, 1988), appears to influence hematopoiesis during fetal life. In adult humans, however, the presence of the Type 5 tartrate–resistant acid phosphatase, which has high amino acid sequence homology with uteroferrin, is indicative of abnormal function of cells associated with hematopoietic tissues (Ketcham et al., 1985) in such diseases as hairy cell leukemia. It is not clear why there is such abundant secretion of uteroferrin by pig uterus and human term–placenta during the course of a normal pregnancy while the apparently identical protein is associated with pathological conditions in adults, e.g., hairy cell leukemia. An understanding of this paradoxical situation requires further studies to determine the precise role of uteroferrin in hematopoiesis. Animal agriculture, veterinary medicine and human medicine will benefit from collaborative efforts to understand the role of this protein in health and disease.

Reproductive Management

Research with farm animals has contributed significantly to the development and use of clinical methods to enhance fertility through the use of techniques for *in vitro* fertilization, embryo transfer, embryo culture, endocrine therapy, cloning of embryos and cryopreservation of sperm, ova and embryos. These techniques are used to enhance fertility in human and nonhuman animals. Methods for contraception or reducing fertility are also based, in large part, on results of experiments with domestic livestock. Functional sterility is desired by many humans to limit family size and in animals it may be used to control growth of populations in general or in a specific geographical area. Research continues to develop methods which render one functionally sterile, but are reversible. Progress in development of immunological methods for achieving fertility control continue and will impact programs designed to offer safe and effective alternatives for regulation of fertility in human and nonhuman animals.

DEVELOPMENT OF ANIMAL GENETIC MODELS OF DISEASE

Development of animal models to study human disease often raises ethical and/or animal welfare issues. However, these issues must be considered in light of the tremendous positive impact disease research has had in the past and will undoubtedly provide for the future for both human and nonhuman animals. Diseases of humans and animals must be understood if the adverse impact of those diseases are to be ameliorated or eliminated. The possibility of experimentation with naturally occurring or transgenic animal models expressing diseases, or metabolic disorders, is essential to the biomedical community. Otherwise, we could not adequately address issues of health, disease and reproductive management that our increasingly global society will present during the next century.

REFERENCES

Bazer, F.W., H.M. Johnson and J.K. Yamamoto. 1989. Effect of ovine trophoblast protein–1 on replication of the feline and human immunodeficiency viruses. *Biol. Reprod.* 40: Suppl 1: 63.

Bazer, F.W. and H.M. Johnson. 1991. Type I conceptus interferons: Maternal recognition of pregnancy signals and potential therapeutic agents. *Am. J. Reprod. Immunol.* 26:19–22.

Bazer, F.W., D. Worthington–White, M.F.V. Fliss and S. Gross. 1991. Uteroferrin: A progesterone–induced hematopoietic growth factor of uterine origin. *Exp. Hematol.* 19:910–915.

DeCicco, F., G. Sica, M.T. Benedetto, G. Ciabattoni, F. Rossiello F., A. Nicosia, G. Lupi, F. Iacopino, S. Mancuso and S. Dell'Acqua. 1988. 'In vitro' effects of β–interferon on steroid receptors and prostaglandin output in human endometrial adenocarcinoma. *J. Steroid Biochem.* 30:359–362.

Haase, A.T. 1986. Pathogenesis of lentivirus infections. *Nature.* 322:130–136.

Janssen, J.L.S., L. Berk, M. Vermeulen and S.W. Schalm. 1990. Seizures associated with low dose α–interferon. *The Lancet.* 336:1580.

Ketcham, C.M., G.A.Baumbach, F.W. Bazer and R. M. Roberts. 1985. Immunological, chemical and enzymatic similarities between uteroferrin and Type 5 acid phosphatase isozyme from human leukemia hairy cells. *J. Biol. Chem.* 260:5768.

Ketcham, C.M. 1988. The human Type 5, tartrate resistant acid posphatase: purification, characterization and molecular cloning. Doctoral Dissertation. University of Florida, Gainesville.

Newton, G.R., J.L. Vallet, P.J. Hansen and F.W. Bazer. 1989. Inhibition of lymphocyte proliferation by ovine trophoblast protein–1 and a high molecular weight glycoprotein produced by preimplantation sheep conceptus. *Am. J. Reprod. Immunol. Microbiol.* 19:99–107.

Oldham, R.K. 1985. Biologicals for cancer treatment: interferons. *Hospital Practice.* 20:71–91.

Ott, T.L., G. Van Heeke, M.H. Johnson and F.W. Bazer. 1991. Cloning and expression in S. cerevisiae of a synthetic gene for the Type I trophoblast interferon ovine trophoblast protein–1: purification and antiviral activity. *J. Interferon Res.* 11:357–364.

Ott, T.L., Y. Zhou, M.A. Mirando, C. Stevens, J.P. Harney, T.F. Ogle and F.W. Bazer. 1992. Changes in progesterone and estrogen messenger ribonucleic acid and protein during maternal recognition of pregnancy in ewes. *J. Mol. Endocrinol.* (submitted).

Pontzer, C.H., F.W. Bazer and H.M. Johnson. 1991. Antiproliferative activity of a pregnancy recognition hormone, ovine trophoblast protein–1. *Cancer Res.* 51:5304–5307.

Tamm, I., B.R. Jasny and L.M. Pfeffer. 1987. Antiproliferative action of interferons. In *Mechanisms of Interferon Actions, Vol II.* L.M. Pfeffer, ed., pp 26–57. CRC Press, Inc., Boca Raton, FL.

Whaley, A.E., R.S. Carroll, K.P. Nephew and K. Imakawa. 1991. Molecular cloning of unique interferons from human placenta. *Biol. Reprod.* 44 (Suppl.1):186.

Lawrence Busch
Sociology
Michigan State University
John J. Kopchick
Edison Animal Biotechnology Center
Ohio University

Workshop Report

Until recently, animal agriculture and human medicine were largely distinct fields of scientific inquiry. However, as a result of the advances made in molecular biology over the last decade, the two fields have come to share a wide range of techniques and models. These profound changes in the research process have raised a series of issues with respect to the use of both farm and traditional laboratory animals in research. The discussions in this workshop session focused on these issues and recommendations were developed.

The links between animal agriculture and human medicine led participants in this workshop to the exploration of various relevant issues. They are discussed in depth below.

HUMAN HEALTH CONSIDERATIONS THAT DRIVE AGRICULTURAL RESEARCH

In recent years, public concern about food safety and nutrition has played an increasing role in animal agricultural research. Often, public concerns and scientific perspectives with respect to these issues have differed. Yet, understanding public concerns and engaging in genuine dialogue are essential if these issues are to be properly addressed.

The public is also concerned about the disclosure of the contents of food and food products. One area of concern is the labeling of foods which is currently being debated both nationally and internationally. In addition, there are public concerns about broader marketing issues, e.g., product claims.

Participants asked how one chooses between the various options for resolving particular problems or improving situations. For example, reduced fat intake could be brought about by: 1. developing lower fat meats through molecular biology; 2. cutting excess fat off on meat; 3. reducing marbling of meat through conventional technologies and changes in the diet of meat animals; or 4. urging consumers to consume less meat than they do currently.

In addition, new biotechnologies blur the lines between nutrition and pharmacy, making possible the creation of what have been variously called "nutraceuticals" and "pharmafoods." These products, often of animal origin, serve a combination of nutritive and therapeutic goals. They raise complex issues of regulation, food safety and consumer education.

ETHICAL USE OF ANIMALS

Some participants at NABC 4 argue that the use of animals as food or in research is itself unethical. Others argue that humane treatment of animals is the major concern. The workshop participants agreed that it was not clear just what is ethical. Moreover, they were concerned with methods used to accommodate the wide range of views on the subject found in our diverse society.

A related area of concern was the extent to which human beings should be dependent on animals for food, medicine and medical research. Some would argue that such dependence should be limited, while others consider it unproblematic.

Some of the concern regarding the ethical use of animals in biotechnological research stems from the public perception that biotechnological research differs from other biological research. Without question, venture capital firms and research universities seeking new funds have fostered that belief. Yet others argue that there are no special concerns about the use of animals arising from biotechnology.

The creation of transgenic animals is a case in point. Some persons argue that this marks a major departure from earlier animal research. Some ask whether it is ethical to transfer genes across species lines in order to serve human ends. The utilitarian position asserts that such research is acceptable as long as the benefits outweigh the costs and risks. Yet some participants suggested that the utilitarian position might have limitations insofar as it produced side effects for animals. A related issue is how much genetic alteration is acceptable and how much is too much.

Finally, participants questioned whether current guidelines on the use of animals in research, often written before the advent of the new technologies, are adequate morally.

ANIMALS FOR BIOLOGICS AND THERAPEUTICS

The use of animals for the production of vaccines and therapeutics has a long history. For example, porcine and bovine insulins, purified from the pancreas of these animals, have been used therapeutically for human diabetes. Nevertheless, the widespread use of animals as living "bioreactors" to produce chemicals of value to humans, poses several important issues.

Most obvious of these is the ethics of use. Put differently, to what extent do we have a moral right to transform animals in this way? Workshop members indicated that this type of use differed from other uses of animals (e.g., in food and fiber production) and were concerned as to what, if any, ethical implications were associated with it.

Moreover, use of animals as bioreactors raises other questions, both ethical and practical. For example, there is the problem of what to do with the carcasses of these animals. Should they be allowed to enter the food chain? Unlike similar products produced through cell or tissue culture, animal bioreactors raise special questions of regulatory clearance and quality control, since cell lines are produced in a controlled environment, whereas animals are in contact with other organisms, including those causing diseases.

Animal bioreactors also pose problems of containment, welfare and management. Such animals probably would have to be segregated from other farm animals, raising the question of whether it would be more desirable to have certain species earmarked for this purpose and not used for food. Moreover, such animals might require greater care or different methods of livestock husbandry than is typical of conventional farm animals.

This, in turn, raised the issue of whether whole animals or cell cultures should be used for screening of therapeutic products. While it was generally felt that cultures were usually more economical, it was also noted that certain cases were likely to require the use of whole animals.

THE SOCIETAL CONTEXT OF SCIENCE SHARED BY AGRICULTURE AND MEDICINE

Research rarely takes place outside a larger social context. That context provides both the limits and opportunities for research. The workshop participants found a number of contextual issues of relevance.

A central issue was how (or whether) to integrate private and public research at the agriculture–medicine interface. Included here were a multitude of issues such as the sharing of information (e.g., genome maps and sequences) and access to funds, materials and processes.

Another key issue was the distributive aspects of this type of research. All research activities tend to disrupt ongoing social and economic arrangements. The group asked who benefits, who pays, and who loses from a research project.

Furthermore, new linkages between the medical and agricultural sciences will be influenced by the current state of food, agricultural and medical policy which will, either through their deliberate intent or by unintended side effects, encourage certain kinds of linkages with research and discourage other kinds. At the same time, the discoveries and inventions stemming from this research will have a considerable impact on food, agricultural and medical policies. *There is a clear need to attempt to foresee some of these impacts now.*

The intentional transfer of genetic information between organisms raises other issues. In particular, although the public may be willing to accept transfers of material across species in general, it may be unwilling to accept the transfer of human genes to animals, or vice versa, for any purposes.

Workshop participants also wondered whether the current institutional structures (especially at universities) were adequate for the new linkages between agriculture and medicine. It was noted that medical and agriculture schools, often even those on the same campus, have little history of cooperation.

There is probably much duplication of effort occurring in medical schools, veterinary medical schools and agriculture schools. Currently, the agriculture and medical research agendas are largely distinct and they are set by different constituencies. There was some concern that we do not know how the agenda is currently set, nor will we know who will control the agenda of newly linked agricultural and medical research programs. One of the proposed solutions to these sorts of problems is technology assessment in various forms. The group felt that it was unclear whether the methods and models of technology assessment were up to the task of charting this new area of research.

Another area of concern was that of public relations, both domestically and internationally. Is there a way in which the international issues could be correlated?

Finally, there was a general consensus that greater public participation in the decision-making process was both necessary and desirable.

RECOMMENDATIONS

1. *Stronger links need to be developed between agricultural and medical research relating to biotechnology. Among mechanisms to do so are centers, incentives for joint programs, funding, etc. This will require further integration and institutionalization of joint agricultural and medical programs. Such linkages will need to include an examination of the ethical, economic, social, institutional and legal ramifications of these changes.*

2. *More resources from molecular biology should be devoted to genome and other research in an attempt to ultimately spare animals from direct use in research. It should be thereby possible to shift largely from whole animal to organ, tissue or cellular systems.*

3. *Explore the moral implications of the use of animals in medical and agricultural research. Issues in the area are currently inadequately examined and thus, there is not yet an adequate moral framework for making decisions about this type of research.*

4. *Provide for education of and dialogue among all the participants in the debate.*

5. *Improve the agenda setting process that insures that resources are properly allocated and that all interested parties are involved in the allocation process.*

6. *Improve the guidelines to aid in determining appropriate circumstances for patenting animals, tissues and cell lines.*

H. Russell Cross
Administrator
Food Safety and Inspection Service
U.S. Department of Agriculture

Food Safety Perspectives on Animal Biotechnology

Having been involved in meat and food animal research for a good part of my career, I am aware of the opportunities biotechnology provides in improving the health of food animals and the safety of the meat supply. As the new administrator of the U.S. Department of Agriculture, Food Safety and Inspection Service (FSIS), I am also aware of the responsibilities we have to ensure the safety of food produced through animal biotechnology. In this paper I will present my perspectives on animal biotechnology by discussing two broad areas. First, the opportunities biotechnology provides in making food safer, and second, the regulatory implications of the new technologies as they relate to the meat and poultry supply.

OPPORTUNITIES TO PRODUCE SAFER FOODS

The Food Safety Inspection Service is a public health agency dedicated to ensuring the safety of the meat and poultry supply. For that reason, any new technology that offers the opportunity to fulfill that mission more effectively and efficiently is of interest to us. While we are addressing the issue of transgenic animals, we are also looking at products such as bioengineered vaccines, bioengineered pharmaceuticals and diagnostic tests. All of these will have a long–term impact on animal health and food safety. Biotechnology offers a number of new tools that will improve the health of food animals. Genetically engineered vaccines may confer immunity more safely and efficaciously than traditional vaccines. With genetically engineered vaccines we can differentiate whether the animal is vaccinated or infected—an important distinction in the fight to control and eradicate animal diseases.

One North Carolina firm recently developed a method of vaccinating chickens inside the shell—even before they hatch. While FSIS does not use these vaccines or regulate them, they affect our mission by improving the health of animals coming to slaughter.

Other tools, such as improved diagnostic tests, are of direct value to us in the meat and poultry inspection program. For instance, researchers with USDA's research arm, the Agricultural Research Service (ARS), are developing a recombinant antigen for the serodiagnosis of bovine cysticercosis. We hope to be able to use this test in the inspection program in the future.

Biotechnology also offers opportunities to improve the microbiological safety of meat and poultry products—our number one priority. In our laboratories DNA probes are being used for detection of *Salmonella* in cooked, ready–to–eat meat and poultry products. We are working to integrate similar DNA probes for *Listeria monocytogenes* and *Campylobacter* into our program in the near future. These tests provide advantages in terms of reducing the time needed to get results and greater specificity in identifying organisms.

Another way to improve microbial safety is by using recombinant DNA to produce a bacteriocin effective against specific foodborne pathogens. The bacteriocin could be added to processed foods to reduce spoilage in a manner similar to the currently approved use of nisin in cheese spreads. Other developments are on the horizon. While we are not currently using these technologies in our microbiology program, certainly the potential is there. For instance, polymerase chain reaction (PCR) technology will allow us to amplify the genetic material from pathogens so we can detect these pathogens without enrichment. This technology will also allow us to look for a specific bacterial genus or species and even a specific virulence gene.

Biosensors are an even newer avenue of biotechnology research. By attaching an antibody, enzyme or nucleic acid to an electrode, these sensors can be used to detect a foodborne pathogen or antibiotic. They have the potential to make shelf life predictions for chilled meat by detecting glucose (an indicator of microbial spoilage flora) at the surface of the meat. While detection is important, the ultimate goal is to prevent contamination in the first place. That is why we hope to see future research directed towards using genetic alteration to produce meat and poultry resistant to pathogenic microbes such as *Salmonella*.

While diagnostic tests that are faster and more effective is a great advantage, producing disease–resistant animals is equally important. For instance, ARS has demonstrated that it is possible to identify swine with a genotype that is resistant to trichinosis. With further research, this genotype could be incorporated into domestic swine populations—conferring trichinosis immunity to all future generations of swine.

At Texas A&M, site of NABC 4, animal geneticists Jerry Taylor and Scott Davis are working on a project funded by the U.S. Agency for International Development to determine if individual genes in goats are associated with resistance to *Haemonchus contortus*, a parasitic disease that affects ruminants throughout the world. While this specific study is more applicable to Third World countries, it certainly has relevance for domestic animal production. If a genetic basis for resistance can be incorporated into livestock production, we can produce healthier animals and reduce the need for animal drugs.

There are also other possible benefits aside from disease resistance that may be realized through genetic modification of animals. Some of these possibilities include:

—animals with leaner meat;

—animals that use feed more efficiently;

—animals with better growth features; and

—animals that manufacture biopharmaceuticals for human or animal therapy.

The potential benefits from genetically modified animals appears to be increasing all the time.

REGULATION OF TRANSGENIC ANIMALS

Certainly, these new products of biotechnology such as vaccines, diagnostic tests and disease–resistant animals interest us as ways to make the meat and poultry supply safer. We also have another role—to ensure that transgenic animals produced through biotechnology are safe for human consumption. For purposes of this paper, transgenic animals are animals whose genetic composition has been changed by introducing selected genes from other sources into the line from which the animal is derived.

Food–producing animals involved in transgenic animal experiments are currently considered experimental under existing FSIS regulations that affect the meat and poultry industries. The regulations define experimental animals as those treated with experimental drugs, chemicals or biologics. We have not yet approved the slaughter of any transgenic animals and are still in the process of developing our policy. Since the field of biotechnology is changing as we speak, we recognize that our regulatory oversight will have to change to keep pace with technological advances.

In the *Federal Register* of June 26, 1986, the USDA, in conjunction with the Office of Science and Technology Policy in the Executive Office of the President, stated the Department's intention to regulate foods produced by new methods, such as recombinant DNA techniques, within the existing statutory and regulatory framework. This policy is in line with President Bush's federal biotechnology policy, announced in February, 1992, which emphasizes that federal oversight should be based on risk, not triggered simply by an innovative technology.

We believe the existing system will work because we plan to regulate the *products* of biotechnology, not the process itself. Our inspection program is now prepared to handle many diverse animals and many different product types. Our system can handle transgenic animals as well. Under our planned regulatory approach, the investigator must specifically request slaughter of any investigational animals involved in transgenic experiments. Whether or not genetic material was successfully incorporated, the following information must be submitted and reviewed by FSIS before the animals are presented for slaughter:

—species;

—genetic changes being attempted or affected;

—technique used to introduce the genetic material;

—results of appropriate scientific methods for detection of the transgene, such as PCR or Southern hybridization; and

—physical condition and appearance of the animal prior to slaughter.

In addition, for animals that have successfully incorporated the genetic material, the following information should also be provided:

—information on the gene product;

—analytical data/results of the gene product analysis; and

—an assessment of animal health and performance, including a veterinarian's observation and examination, and any clinical laboratory data on the overall health of the animal.

If the information meets the criteria under the experimental animal regulations, the animals would be approved for slaughter. A request for slaughter of these animals must be made indicating the location of slaughter. Each animal from transgenic experiments permitted for slaughter would also receive the required antemortem and postmortem inspection by an FSIS inspector and/or veterinarian. This is important because the way in which an animal grows and functions is a reliable indicator that the change was not detrimental to the safety of these animals. Since we will know ahead of time that the trans-genic animals are to be presented for slaughter, we will have the opportunity to examine their growth and general health before they reach the slaughterhouse.

INTERACTION WITH OTHER AGENCIES

In evaluating the food safety of transgenic animals, we would consult with the Food and Drug Administration (FDA), USDA's Animal and Plant Health Inspection Service (APHIS), or the Environmental Protection Agency (EPA) before making a food safety decision.

The FDA is responsible for assuring food from species other than those inspected by FSIS is safe. They are also responsible for assuring that animal drugs are safe, effective and properly labeled, particularly with regard to the safety of residues remaining in the animal at slaughter. The FDA, along with FSIS, is charged with assuring that food additives added to meat and poultry products are safe for consumers. The FDA, in cooperation with state authorities, also sets standards for the wholesomeness of milk. Pesticide chemicals, used directly on food animals or on animal feed crops, are reviewed prior to marketing for safety by the EPA. Finally, biologic products, such as vaccines and serums used in animal health programs, are subject to oversight by APHIS for potential food safety impacts.

To repeat, FSIS has not approved any transgenic animals for slaughter yet. Our policy on these animals is still being developed and will be considered ready for review as soon as FSIS has come to an understanding with FDA regarding jurisdictional responsibilities of the two agencies with regard to animal biotechnology. FSIS is planning to publish a paper entitled "Points to Consider" by the end of 1992 that will offer more specific guidance on the requirements for slaughter of transgenic animals. In addition, we plan to have our entire policy reviewed by USDA's Agricultural Biotechnology Research Advisory Committee (ABRAC).

I also want to emphasize that all federal agencies involved in regulating biotechnology are coordinating their efforts in order to secure common guidelines and a clear understanding of jurisdictional responsibilities. In the near future evidence of this coordination as policies on various biotechnology products emerge.

REGULATION OF NONTRANSGENIC ANIMALS FROM BIOTECHNOLOGY RESEARCH

Although no transgenic animals have been approved for slaughter yet, FSIS has authorized the slaughter of nontransgenic animals in Texas. These animals were involved in biotechnology experiments, but they were not genetically modified. These animals were slaughtered after it was ascertained that the criteria announced in the *Federal Register* notice of December 27, 1991, "Livestock and Poultry Connected with Biotechnology Research" (Vol 56, No. 249) was met.

OBSTACLES TO THE USE OF BIOTECHNOLOGY IN ANIMAL AGRICULTURE

Progress has been made in the animal biotechnology arena and the benefits to animal health and food safety are evident. It will not be smooth sailing all the way, however. There are potential obstacles out there that must be brought into the open in order to address them in a constructive manner.

Consumer acceptance of the new technology is a prime example. Just because it is good technology does not mean consumers will accept it. All of us—government, academia and private industry—must work together to address consumer concerns. At FSIS, better communication with the public about biotechnology as well as all other issues concerning food safety, is one of my major priorities. We must not wait until the questions are asked before we provide information. We must not wait until we are attacked to respond. We must be on the offense, not on the defense.

At USDA, we are developing a strategy to get information about biotechnology to the public with the goal of helping the public make informed decisions about the products of biotechnology. Certainly, our agency will have a role in informing the public about our regulatory strategy regarding transgenic animals, but this is just a small part. We must do much more. This is especially important because we will be competing with a number of other groups for the public's attention on this issue.

Short–term, we must focus on educating U.S. policymakers about biotechnology so they can make informed decisions on legislation and policies at the local level. USDA is partially funding pilot studies currently underway to educate local county administrators on the risks and benefits of biotechnology.

Long–term, we must reach the public. USDA's Extension Service plans to set up focus groups with consumers to determine what types of information the public wants and how best to provide them with that information. We must know our audience and we must know *how* to reach them.

The bottom line is this: we must stay in tune with public opinion. I urge you to pay close attention to a survey of consumer attitudes about biotechnology to be released shortly. It was conducted by North Carolina State University and Colorado State University and funded by USDA's Extension Service and North Carolina State.

While the preliminary results show overall support for the use of biotechnology in agriculture and food production, apparently the acceptability of biotechnology will vary with the specific use. People are much more comfortable with the idea of tinkering with plants than with animals, a reflection of public concerns regarding the well–being of animals and moral beliefs regarding genetic modifications in animals.

It is also apparent that the public wants to be involved in decision–making about biotechnology. This interest is a good sign that the public will be receptive to biotechnology education. That is one reason I have been so candid here. Not only must the public be enlightened to enable them to make informed decisions about biotechnology, but they must have confidence in the government's ability to regulate biotechnology. The public must believe us when we say these products are safe. If we do not have their confidence, use of the technology is threatened. That is why we must carefully develop our policies and involve the public in the decision–making process.

SUMMARY

In summary, biotechnology offers us many opportunities to improve agriculture. I believe biotechnology will have its greatest impact on meat and poultry safety in two ways. First, it will provide us with diagnostic tests that can help us to quickly and effectively detect contamination during the food production process.

Second, biotechnology will enable the production of healthier animals through improved vaccines, improved diagnostic tests and the ability to produce disease–resistant animals. Biotechnology will also potentially provide us with animals with leaner meat, animals that are more feed efficient and animals with better growth potential.

While ensuring the safety of transgenic animals will have an impact on FSIS, I am confident our regulatory structure is equipped to ensure the safety of these new animals.

David Berkowitz
Senior Staff Member
Office of Biotechnology
U.S. Food and Drug Administration

The Food Safety of Transgenic Animals[*]

In the last 40 years a number of modern techniques for improving animal lines have been developed. Artificial insemination has already had an enormous impact on the dairy industry. Techniques such as *in vitro* fertilization, embryo cloning, nuclear transplantation and transgenesis are reaching maturity. These techniques and their potential effects on the environment, genetic diversity, animal production and society have been discussed by George Seidel (1989; 1991).

Introducing food products into the market place requires that the safety aspect be fully analyzed and documented so that healthy transgenic animals will be at least as safe as the traditional animals from which they were derived.

The classical breeding of familiar food animals has been practiced since antiquity and has never resulted in a hereditary trait that made animals unsafe as food. Traditional breeding is accomplished by focusing on a desirable trait, such as milk production or fat content, and breeding only those animals which best exemplify the trait. If the trait is quantitative, this practice moves the population mean in the desired direction. The cause of the improvement is unknown. The progeny results from thousands of selections between paternal and maternal genes and the genes responsible for the improvement in the phenotype are rarely, if ever, identified. There is little knowledge of the physiological mechanism underlying the phenotypic change. Yet, this approach has been safe and successful and is exemplified by the dairy industry where selecting semen from bulls with high–producing daughters more than doubled the milk output per cow in the twenty years following 1955. The genetic events associated with traditional breeding are safe; consequently, only the unique features of transgenesis are examined here.

One can organize the unique features of transgenesis into three categories: the genetic construct (i.e., the DNA introduced), changes resulting from the integration of the construct, and the nature of the gene product.

* The views expressed are not necessarily those of the Food and Drug Administration.

SAFETY CONSIDERATIONS ASSOCIATED WITH THE GENETIC CONSTRUCT

There is little concern about the safety of orally consumed genes. The human diet, consisting of bacterial, animal and plant products, include all of the genetic material of those organisms. Digestive enzymes in the human gastrointestinal tract degrade DNA in the food and, since a single nick in a gene is enough to inactive the production of the gene product, the probability of a functional gene sequence surviving intestinal digestion may be considered near zero. On the off chance that some DNA does survive, it would only be excreted.

The increased purine and pyrimidine content of tissues resulting from the extra gene in transgenic animals will be negligible relative to the total tissue purine content. In mammals, the purine from a single gene is on the order of one millionth of the total genomic content of purines. Some plant breeds produced by traditional methods have resulted in large percentage increases in the nucleic acid content, i.e., increases in the somatic cell chromosome number. These considerations may be more important if the food product were a sole source of protein or energy.

The DNA of the construct is of concern only if it is infectious, i.e., if it can be propagated in the environment or transmitted by the food to susceptible cells in the gastrointestinal tract. Retroviruses are used to introduce genes into some species, particularly poultry. The viruses that come in contact with prospective transgene recipients are defective, likely carrying at least one deletion in a transacting gene. Rarely, through recombination with endogenous viruses or from functional retroviruses present in nature, could fully functional viruses emerge from the helper cell line. The probability of functional recombinants arising is small, but they have been observed. New helper cell lines with less homology between the defective viruses and the provirus will reduce the possibility of recombinant virus production (Miller, 1990; Temin, 1989). From the food safety perspective, even competent animal retroviruses pose no threat to human health because of the species specificity of viral infection.

SAFETY CONSIDERATIONS ASSOCIATED WITH INTEGRATION OF THE CONSTRUCT

The insertion of a transgene into a recipient genome is a safety consideration because the location and manner of insertion may increase or decrease the expression of host genes. The hypothesis is that the insertion process might activate latent toxin genes or increase levels of hormones or other substances detrimental to human health when the food is eaten. In healthy animals this is not a realistic concern. If the transgenic animal is not healthy, the cause must be investigated to be certain that the pathology has not resulted from something transmissible in the food. However, the possibility of activating a toxin gene is insignificant, as discussed below.

The genetic events causing the modulation of gene expression as a result of transgene insertion are not different from genetic events that occur naturally. Modifications of gene expression are caused by the generation of new connections between sequences that are not normally juxtaposed or by the separation of normally connected sequences. Chromosomal translocations, deletions and inversions occur continually in animals in nature as well as in food animals. Animals also contain interspersed sequences that transpose to new chromosomal locations, though the frequencies of transposition in food animals are not known. Written records of animal breeding go back as far as Aristotle (Sturtevant, 1965), and animal breeding has never been associated with the production of toxic lines of animals. This historical record is strong evidence for the food safety of translocations, inversions, deletions and insertions in animal chromosomes.

Toxin genes are rare in animals. Animals are generally safe as food. The dangers in eating animal products usually stem from parasites or microbiological contaminants; these are inactivated by cooking. The overwhelming majority of animal species can be eaten without harm. There have been reports of dogs being poisoned by eating polar bear liver, but the poisonings are caused by high levels of vitamin A in the livers (Russel, 1966). Although this is an example of toxicity from the ingestion of animal tissues, the accumulations of high levels of vitamin A in the liver is a complex trait and is not induced by a single genetic event. The genomes of the common food animals do not carry toxin genes that can be activated.

A classical case of acute "animal" poisoning is the biblical case of quail poisoning (brought to my attention by John Kirschman) described in *Numbers, Chapter 11*. During the Exodus the Israelites became tired of eating manna and wanted "flesh."

> And there went forth a wind from the Lord, and brought across quails from the sea...and the people gathered quail; he that gathered the least gathered ten heaps...While the flesh was yet between their teeth, ere it was chewed, the anger of the Lord was kindled against the people, and the Lord smote the people with a very great plague. And...they buried the people that lusted.

The investigation of modern cases of quail poisoning have been attributed to coniine, the hemlock neurotoxin that killed Socrates. The quail feed on hemlock during their migration from Africa to Europe, are resistant to the toxin, and are able to consume enough hemlock to poison predators. The toxin itself is a plant product, not an animal product.

An important consideration in animals is that toxic genetic effects with adverse human health effects, unexpected or otherwise, are likely to produce visible signs in the development or growth of the transgenic animal. Transgenic animals are themselves an important demonstration of their food safety. The fact that an animal has gone through normal intrauterine development, birth and growth in the presence of the transgene and its product is a strong indication of the safety of the derived food. For the food to be toxic, the animal would have to produce a species–specific toxin that is inactive in the species of origin, but orally active in the species consuming the food. No such toxins from land food–animals have been described.

THE SAFETY OF THE GENE PRODUCT

The essence of the safety review of transgenic animals must be an examination of the gene product. The safety of gene products may be reviewed in the same way the safety of drugs or pesticides are classically reviewed, i.e., the important food safety matter is the presence of a pharmacologically or toxicologically active residue. Because the product of the transgene is completely characterized, one can use traditional methods to evaluate its safety. This is an advantage over traditional breeding because the knowledge of the exact genetic change directs the safety inquiry to the correct gene product and its effects. Traditional breeding is accomplished empirically by focusing on a desirable trait with little knowledge of the physiological mechanism underlying the phenotypic change.

Gene products may have both direct effects resulting from the action of the gene products themselves and indirect, secondary or compensatory effects brought about in response to the direct effects of the gene product. For example, somatotropin stimulates the secretion of IGF–1 from the liver and other tissues. IGF–1 is responsible for many of the effects formerly attributed directly to growth hormone and this was taken into account in evaluating the safety of milk from bovine somatotropin–treated cows (Juskevich and Guyer, 1990). Such reasoning is normally part of the review of the food safety of feed additives and new animal drugs. Routine toxicology testing is designed to detect all effects of a compound, direct and indirect.

Once the safety of the transgene product is established, transgenic animals may be considered as safe as traditional animals. Some of the food safety considerations may change as the technology advances. Richa and Lo (1989) produced "transomic" mice by introducing chromosome fragments dissected from metaphase spreads into fertilized ova. Chromosome fragments known to be associated with desired traits can be used selectively. Large numbers of genes (10 megabases) are introduced rather than selected genes. For intraspecies transfers, the results are likely to be similar to naturally–occurring cases of trisomy. Transomic animals are likely to be safe also, but too few have been studied to make conclusions about the food safety considerations.

If we imagine that we are many years in the future when livestock are routinely improved by recombinant DNA techniques, traditional breeding, in retrospect, will seem far too hazardous. To allow all the genetic changes to occur by chance and then never know what genes or genetic changes were responsible for the new phenotypes is likely to seem far more risky than transgenesis. Cattle have 30 pairs of chromosomes. Thus, *in the absence of recombination,* a single mating has a potential of producing 2^{30} or 1.07 billion genetically different eggs or sperm. Surely the introduction of a single well–characterized known gene is less risky!

REFERENCES

Juskevich, J.C. and C.G. Guyer. 1990. Bovine growth hormone: human food safety evaluation. *Science.* 249:875–884.

Miller, A. D. 1990. Retrovirus Packaging Cells. *Hum. Gene Therapy.* 1:5–14.

Richa, J. and C.W. Lo. 1989. Introduction of Human DNA into Mouse Eggs by Injection of Dissected Chromosme Fragments. *Science.* 245:175–177.

Russel, F.E. 1966. Vitamin A Content of Polar Bear Liver. *Toxicon.* 5:61–62.

Seidel Jr., G. E. 1989. Genetics in the Pasture. *Technology Review.* 92:42–52.

Seidel Jr., G. E. 1991. Biotechnology in Animal Agriculture. In *NABC Report 3, Agricultural Biotechnology at the Crossroads: Biological, Social and Institutional Concerns.* J.Fessenden MacDonald, ed. National Agricultural Biotechnology Council. Ithaca, NY p. 97–108.

Sturtevant, A.H. 1965. *A History of Genetics.* Harper and Row, New York.

Temin, H.M. 1989. Retrovirus vectors: Promise and reality. *Science.* 246:983.

Dianna Hunter
Writer
Member, Minnesota Food Association

To Live as Natives, Free of Fear: What Citizens Should Require from Animal Biotechnology

Floodwood, Minnesota, the little town near my farm, can barely keep a feed store going today, but it led the nation in dairy technology 55 years ago. In 1937 at Island Farm, on the flat marsh west of town, technicians performed the first-ever insemination of a dairy cow with frozen semen (Goodrich, 1988). Into the early 1960s, the roads around Floodwood were lined with dairy farms and Floodwood had three car dealers, two tractor dealers and a cooperative creamery owned by local farmers. Today, we have to drive 40 miles to buy a new car or tractor and the only cooperative creamery left in all of northeastern Minnesota is the Duluth Division of Associated Milk Producers, Incorporated—a cooperative owned by farmers from Minnesota to Texas. During the Reagan years, half the farmers who shipped milk to that last local creamery quit dairying (Hunter, 1989).

Considering America's agricultural history, it is not surprising that farmers and other citizens who follow agricultural events ask hard questions and look with cynicism at the promises made for genetically engineered animal products such as bovine growth hormone and porcine growth hormone, and humanly created, patented species. To Floodwood farmers, biotechnology looks like one more unit in a long parade of agricultural technologies—technologies which were sold to us as benefits, but which led to the displacement of our neighbors and the decline of our towns.

We are wary of new technologies because of our experiences. Besides the hurt heaped on family farmers by technology-driven economic upheaval, we have witnessed the hurt heaped on everyone by technology-driven environmental upheaval. Jim Davidson, soil scientist and research dean at the University of Florida, did a good job of articulating the reasons nonfarming citizens are wary in his 1989 address to the Agronomy Administrator's Round table. Davidson said:

> The distrust on the part of nonagricultural groups is well justified. With the publication of Rachel Carson's book entitled *Silent Spring*,

> we in agriculture loudly and in unison stated that pesticides did not contaminate the environment. *We now admit that they do.* When confronted with the presence of nitrates in groundwater we responded that it was not possible for nitrates from commercial fertilizer to reach groundwater in excess of 10 parts per million under normal productive agricultural systems. *We now admit that they do.* When questioned about the presence of pesticides in food and food quality, we assured the public that if a pesticide was applied in compliance with the label, agricultural products would be free of pesticides. *We now admit that they're not* (Kirschenmann, 1992).

Since informed citizens have such good reasons to be wary of promises made for new technologies, it is a challenge to find avenues of communication between them and the proponents of animal product biotechnologies. My job in this essay is to suggest some possibilities. I am not going to dwell on the bovine growth hormone (BGH) experience, but I *am* going to use it as a springboard, an element in a true story that illustrates why citizens have learned to require honesty, patience and respect from the proponents of new animal biotechnologies.

As a dairy farmer in the mid1980s I started following news about BGH in farm magazines. In 1988, I joined a year–long biotechnology study group through a membership organization called the Minnesota Food Association. In 1990, after I had sold my cows, a farm woman who sits on the advisory board of our local agricultural experiment station called and asked me to attend the station's meeting on BGH. "I know you'll ask good questions," she said. I called another friend, a young woman who is taking over her father's dairy farm, and on a cold day in January, we met about halfway to Grand Rapids, at the Swan River truck stop, and drove the next twenty miles together.

In a basement room, two animal science researchers from the University of Minnesota presented a six–hour lecture program on the hormone they called "BST" (bovine somatotrophin). During the morning, I took notes and asked a few questions including who had provided funding for the research. The public had provided some money, it turned out, but most had come from Monsanto and American Cyanamid, two pharmaceutical companies that planned to market BGH/BST. At the end of the morning session, one of the researchers waved me over and asked whether I meant to imply that he was "in the pocket" of the pharmaceutical companies. I told him it was not that simple in my mind, but I thought that we all ought to consider what it means when universities choose research projects based of the amount of money the research can solicit from private industry. We had a long, friendly discussion which cut into the lunch hour.

My friend and I were pushing plastic trays down the cafeteria line when the other researcher approached us and spat, in an exasperated voice, "What is it that you're afraid of?" Caught off guard, I spat back something like, "The destruction of our farms and our communities." I meant to go on, hoping to make the uncomfortable exchange evolve into a more civil one, but when I paused for breath, the man twirled on his heel and hurried away. I felt my face flush with insult.

In the pickup on the way home, I thought about his question. He had misused it by trying to intimidate me, but it was a surprisingly deep question. It occurred to me that we ought to examine our fears more often. Fears are not just embarrassing details; they are essential pieces of human equipment. What other basis do we have for respectful, careful deliberation in the face of danger? How else can we define safety, except to say that it means the absence of good honest reasons for fear? Everyone seems to agree that safety is the one thing which citizens have a right to expect from products of animal biotechnology. The researcher's question was right on the mark: ***What is it that I am afraid of?***

First, I am afraid that animal biotechnologies will be just like other agricultural technologies, pushing along existing trends that benefit agribusiness industries but damage the environment, farmers and rural communities. I am afraid animal biotechnologies may be piecemeal solutions that do not take into account ecological or social systems. I am afraid they will decrease normal, healthy variation within and among breeds. I am afraid they will hurt people in the Third World whose economies are already threatened by genetically engineered plant products such as sugar substitutes, vanilla flavoring and cocoa butter (Jamal, 1988).

I am even more afraid that animal product biotechnologies will be unlike previously known technologies. I am afraid of disasters like the 1989 L–tryptophan poisoning which so far has left 31 Americans dead and 1,500 sick from a blood disease linked to a mysterious double molecule in a genetically engineered food supplement (Raphals, 1990a,b,c; National Wildlife Federation, 1990). If we know all we need to know about the safety of genetic engineering, it is hard for me to understand what went wrong in the L–tryptophan incident. I am also afraid of monster animals like the giant cloned-calves that could not be born vaginally, a failed experiment that drove Grenada Biosciences of Houston into receivership (Hodgkinson, 1992).

I am afraid of one more thing: human ignorance. Technologies are not inherently evil, but if recent history is any guide (See Dean Davidson's list above for just three examples), we humans are not yet sophisticated enough to predict the impact of singular changes on large, interconnected systems. My own particular ignorance scares me, too. It may be that biotechnology is "going to require more of us as citizens than we can handle," as Kansas geneticist, Wes Jackson, predicted (Eisenberg, 1989). The issues are so complex —not just scientific and technical, but ecological, ethical, economic, social and political. To even begin to understand them, ordinary citizens need a crash course. Fortunately, some of us have been able to get that from nonprofit organizations like the Minnesota Food Association, the National Wildlife Federation and the Rural Advancement Fund International.

The Minnesota Food Association's biotechnology study group, in which I participated, makes an effective model for citizen study groups. The Association is a membership organization of people interested in food and agricultural issues. Members identified biotechnology as an important upcoming issue in 1987, and decided to convene a study group. Association staff provided research and support, and they organized a series of informational meetings between study group members and various experts. Study group members included farmers, writers, a veterinarian, a biologist and a futurist. Experts included legislators, ecologists, ethicists, scientists, university administrators and representatives of businesses involved in biotechnology. The group met twice a month and heard from three experts at each meeting. After a year of study (including the meetings as well as the reading of relevant materials) staff and members collaborated to write and publish a report, *Food and Agricultural Biotechnology in Minnesota: A Citizens' Perspective* (Minnesota Food Association, 1988). The report helped spur the Minnesota Legislature into passing one of only two state laws in the country that regulate environmental releases of genetically engineered organisms.

We heard later that some of the scientists who met with the study group felt affronted to have their research questioned by nonscientists. Some of the nonscientists felt affronted by what they judged to be condescension from some of the scientists. No one yelled or was injured. We can bear such small, nonviolent discomforts. They are prerequisites for real communication among equals who do not necessarily agree.

In contrast, one–sided events, like the bovine growth hormone meeting described above, are not real communication. Today's wary citizens know that. We have learned something from three decades of watching television commercials. We know that real communication is not a one–way street, not "reaching" someone with a message, the way public relations firms try to do. Real communication takes place between equals at an intersection with many points of view and many ways to go. Only through real communication can anyone hope to convince us that a product of animal biotechnology is safe—if, in fact, it is.

Consider the bovine growth hormone experience. Neither farmers nor consumers asked for BGH in the first place. That was the first mistake—to develop a product that met no clearly defined need. Neither farmers nor milk drinkers wanted it. Forty–six percent of Minnesota farmers have said they would never use it (Crooker and Otterby, 1990). Eighty–two percent of rural nonfarm North Carolina residents said they were very concerned or somewhat concerned about it (Sorenson, 1990). BGH is a textbook example of how not to develop a technology. Let us learn something.

When considering new genetically engineered animal products in the future, we should ask these fundamental questions: Who wants the technology that might result from this research? Do we hear citizens asking for this technology or will we have to try to manufacture their need for it through one–sided promotional events? Who will profit? Who will pay?

Given current citizen activism and wariness, no one should expect to develop a new technology without public comment, particularly if public funds are involved and if the research is being conducted at land grant universities that have inherent public interests. One thing is clear from the Minnesota experience: the nature of the public comment is negotiable. People can dialogue with structure and moderation, hearing many voices or they can monologue in strife and chaos, employing secrecy, name–calling, moratoriums and protracted battles over legislation and regulation.

Again, the Minnesota Food Association provides a model. In February, 1992, members entered into a moderated dialogue with Gene Allen, a University of Minnesota Vice President who had signed testimony opposing regulations that would implement Minnesota's biotechnology law. This was the same law that the Minnesota Food Association's biotechnology report had helped to pass three years earlier. People on both sides were apprehensive going into the dialogue, but coming out, Dean Allen quipped that, "We are formed by those whom we meet with, and thank God, I don't meet only with vice presidents." (Northern Tier Land Grant Accountability Project, 1992) Further, he invited members of the Minnesota Food Association to meet with the University's Council of Biological Deans who, he allowed, were better suited to answer the members' questions.

I hope that other proponents of biotechnology will open themselves to dialogue with citizens who have joined membership organizations and taken the trouble to educate themselves about biotechnology. Such dialogue is not just public spirited, it is also practical. Citizens already involved in the biotechnology issue are the ones who are likely to cause trouble in the future should some technology look unsafe or ethically cloudy. By hearing their concerns in advance, biotechnology proponents can head off future disagreements like the ones that have hobbled BGH.

Industry executives could also head off future trouble by implementing real communication. Instead, they are being secretive—withholding data that the companies judge to be "Confidential Business Information." Take the case of Frito–Lay, the potato chip company. Under the Freedom of Information Act, the National Wildlife Federation (NWF) has obtained nine of Frito–Lay's USDA applications for environmental releases of genetically engineered organisms. In six of those applications, according to the NWF, Frito–Lay withheld the identity of added genes and other information needed to assess environmental risks involved in the releases (National Wildlife Federation, 1992).

Citizens are tired of being kept in the dark and we do not want to "be reached" by proponents of biotechnologies before we have a chance to say whether we need the technologies or their products. Right from the start, we want a chance to ask questions and express our fears. We want to know if a new technology is going to be like every other technology—if it is going to add to corporate balance sheets and subtract from the balance sheets of American farmers and Third World citizens. We want to know if the new technology is going to be unlike every other technology—if it has the potential to damage society or the environment in unforeseen ways. We want to know how we can educate ourselves so that we can participate as equals in political decisions being made about biotechnology. We want to know that people in the universities are there to listen and to help us get truly educated and that they are not trying to sell us technologies or products.

Minnesota writer and ethicist, Carol Bly, set a practical benchmark in her foreword to my oral history collection, *Breaking Hard Ground* (Hunter, 1991). She wrote, "What we all want is a world in which small operators who like their work can live without any insult and injustice, can live in the places which are native to them and can consort with those they do business with without fear."

With that ethic in mind, let us respect one another, study together and take all the time we need to arrive at careful decisions. Let us not give in to the profit–driven rush to develop genetically engineered animal products. We humans have been practicing animal husbandry for at least 10,000 years (Lerner, 1986). We are not likely to hurt ourselves if we take another 10 or 15 years to carefully test and deliberately study a new product. We might even move a step or two up the evolutionary ladder if we learn to manage a prolonged, civil discourse among disagreeing parties.

REFERENCES

Crooker, B. and D. Otterby. 1990. Unpublished comments citing a University of Minnesota Economics Department poll.

Eisenberg, E. 1989. Back to Eden. *The Atlantic.* 264:57–59.

Goodrich, D. 1988. Unpublished interview.

Hodgkinson, N. 1992. *Sunday London Times.* March 15.

Hunter, D. 1989. Renewing Northeastern Minnesota Agriculture. *New North Times.* Grand Rapids, MN.

Hunter, D. 1991. *Breaking Hard Ground: Stories of the Minnesota Farm Advocates.* Holy Cow! Press, Duluth, MN.

Jamal, A. H. 1988. *The Socioeconomic Impact of New Biotechnologies in the Third World, Development Dialogue.* Dag Hammarskjold Foundation, Uppsala, Sweden.

Kirschenmann, F. 1992. *Biotechnology and Sustainable Agriculture.* Northern Tier Land Grant Acccountability Project, St. Paul, MN. citing John

Pesek's February, 1990, unpublished paper, Research Findings on Alternative Methods.

Lerner, G. 1986. *The Creation of Patriarchy.* OxfordUniversity Press. New York.

Minnesota Food Association. 1988. *Food and Agricultural Biotechnology in Minnesota: A Citizens' Perspective.* St. Paul, MN.

National Wildlife Federation. 1990. Tryptophan: Biotechnology's First Fatal Mistake? *The Gene Exchange.* I(3-4):1.Washington, D.C.

National Wildlife Federation. 1992. The Closing Door. *The Gene Exchange.* III(1):8.

Northern Tier Land Grant Accountability Project. 1992. Reports from the States. *At the Crossroads: Newsletter of the Northern Tier Land Grant Accountability Project.* St. Paul, MN.

Raphals, P. 1990a. Disease Puzzle Nears Solution. *Science.* 249:619.

Raphals, P. 1990b. L–Trytophan Puzzle Takes New Twist. *Science.* 249:988.

Raphals, P. 1990c. Does Medical Mystery Threaten Biotech? *Science.* 250:619.

Sorenson, A.A. 1990. Farmer Concerns: Food, Safety, Biotechnology and the Consumer. In *NABC Report 2, Agricultural Biotechnology, Food Safety and Nutritional Quality for the Consumer.* J. Fessenden MacDonald, ed. National Agricultural Biotechnology Council. Ithaca, NY p. 103–117.

Susan K. Harlander
Director, Dairy Foods Research and Development
*Land O'Lakes Inc.**
Kate Clancy
Nutrition and Food Management
Syracuse University

Workshop Report

To set the stage for further discussions, the workshop began with presentations from two speakers who offered alternative viewpoints on the impact of biotechnology on meat and animal product safety. The view of the first speaker, David Berkowitz, Office of Biotechnology, Food and Drug Administration, was that healthy transgenic animals are as safe as traditionally bred animals if the transgene product is safe. Biotechnology provides the potential to predict, understand and control the genetic basis of animal improvement in precise ways.

The perspective of a rural resident, former farmer and member of the Minnesota Food Association, a nonprofit organization interested in food and agricultural issues, was provided by Dianna Hunter. She broadened the definition of safety beyond animal and meat product safety in the marketplace. Ms. Hunter defined safety as "the absence of good honest reasons for fear" and for her, there were many reasons to fear animal biotechnology.

Ms. Hunter also warned against public relations–style communication models which seek to tell in monologue rather than to listen in dialogue. She reinforced the need for dialogue between groups representing divergent views about biotechnology with open and honest communication and mutual respect for alternative viewpoints.

A pre–meeting survey of registrants found over 80 percent of respondents disagreeing that foods derived from the products of agriculture biotechnology will be less safe than today's food. However, workshop participants, after review, did identify some potential safety problems for discussion. These included unanswered questions about bovine somatotropin (BST), allergenicity and questions about a number of products for which there are as yet no data bases, for example, transgenic animals and animals administered recombinant DNA products. On the positive side, the participants acknowledged the promise identified by past NABC attendees for new biotechnologies to produce diagnostic tools for food safety testing of animal products (See *NABC Report 2,* 1990).

---

*When this workshop was held, Dr. Harlander was with the University of Minnesota, Department of Food Science and Nutrition.

Finding common ground was more difficult and frustrating once the group moved past the fairly narrow, but controllable, technical hazards to the myriad of intellectual and social elements that people bring to a decision about the safety of any entity, food included. Before moving to the identification of social issues, participants identified elements from three other categories—animal welfare, the environment and social concerns. The first was the topic of another workshop and was not pursued further. Some ecological/environmental problems were mentioned including those arising from the release of transgenic fish, the possible narrowing of the genetic base for domestic animals and the, as yet, unstudied effects of the "short–circuiting" of adaptation in domestic species of animals through genetic engineering.

At this point, participants stepped back to list the major concern of each of the participants about the safety of biotechnologically produced meat and animal products. The items fell into four different areas, each listed and discussed below, including two that had significant social aspects. Small groups were formed to discuss these issues and bring recommendations back to the total workshop group for discussion.

THE SAFETY OF TRANSGENIC ANIMALS AND ANIMALS ADMINISTERED RECOMBINANT DNA PRODUCTS

—*Use of transgenic animals to produce pharmaceutical agents for use by humans.* Certain transgenic animals are producing pharmaceuticals for use by humans, such as pigs producing human hemoglobin and sheep producing a blood clotting factor. These "pharm" animals may enter the human food supply, but before they do, their safety must be assured. ***All workshop participants agreed to the need for a data base on the nutrient composition and levels of relevant hormones and residues in these animals to reassure scientists and the public that there are no detectable differences from levels of these substances in traditional animal products.*** There was not a consensus in the group as to how extensive the data base would be and what it would contain.

—*Animals administered recombinant DNA products:* 1. hormones—there are provisions for their regulation by FDA already in place (i.e., regulation of recombinant BST); 2. vaccines of three types—inactivated, gene deletions and live–vectored. The latter two are the ones of concern. None are licensed for release yet although one is being field tested. ***There was consensus that the regulations under the National Environmental Protection Act (NEPA) and the testing protocols were probably adequate***; and 3. direct–fed microbials—these are feed additives such as yeasts, bacterial enzymes and probiotics. FDA has the authority to regulate these but has not been doing so. ***Participants agreed that FDA should investigate direct–fed microbials more carefully in the future, when applications for recombinant products are received.***

—*Long–term consequences of breeding transgenic animals.* The concern here is the unknown potential for unexpressed genes to cause other changes in animals that may not be expressed for several generations. Some in the group believed that animals should be observed for longer than one generation to detect any such changes. Others believed that observation of the first generation of offspring was sufficient. The group did not agree on whether other data bases should be developed on transgenic animals to assure the public that there are no differences in the levels of various chemical compounds in the meat of these breeds compared to animals now on the market. ***The final recommendation in this area speaks to the need for remaining aware of the possibility of cloning defects in embryo transfer and cloning experiments.***

BIOTECHNOLOGICAL TOOLS TO ENHANCE FOOD SAFETY AND QUALITY

—*Animal products are the major source of microbial contamination in the food supply, so that use of DNA probe assays and immunoassays for the detection of pathogens is to be strongly encouraged.* Large–scale detection of pathogens is impractical with present technology. Unavailability of rapid and economically effective methods for detection of undesirable materials and contaminants during animal production and processing hinders application of intensive inspection protocols. Biotechnology is the most promising source of tools that can yield rapid, sensitive, specific and cost–effective diagnostic tests for the presence of microbiological pathogens, antigens, toxins and other compounds–of–interest to improve food safety. New diagnostic capabilities can also be used to detect adulterated foods and as a screening method for allergens in the food supply.

The rapid detection of contaminants should lead to the development of improved processing methods and a decrease in the incidence of food borne illness in the population. ***There was consensus in the workshop that research and application of these tools should move ahead rapidly.***

—*Genetic markers also offer the potential to improve the healthfulness and safety of the food supply.* They allow more rapid and effective application of traditional or conventional genetic selection practices. These new techniques can improve selection for multiple beneficial traits without a substantial loss of progress in other traits of interest. Improvement of resistance to diseases, or colonizations by parasites or human pathogens, decrease the frequency of application of therapeutic drugs and moderate degeneration of animal health, thereby reducing the presence of unwholesome products in the food supply. Genetic markers can also be used to breed for improved macronutrient composition, such as decreased fat in animals. ***For these reasons the group also endorsed continued research on the use of the genetic markers techniques.***

DEFINING FOOD SAFETY

—***The larger issue here is how to define food safety.*** Some participants argued the present definition is too narrow, ignoring quality issues as well as the fact that food safety is a social construct, as illustrated by the different definitions and standards for food safety held by different cultures and countries. They felt that social, economic and political issues should be evaluated concurrently with the evaluation of efficacy and human and animal safety. Others disagreed with all of these ideas and argued for maintaining the present system of relying solely on technical data for safety decisions. The latter participants did recognize that social, economic and political issues should be discussed. After further comments the workshop debated a recommendation that a mechanism should be set up for formal consideration of the social, economic and environmental ramifications of agricultural biotechnology products. It was noted that there are already regulatory requirements for reviewing environmental consequences, but the group felt it important to state the need for review of environmental consequences. There was not consensus about whether the mechanism should be separate from, or integral to, the present system.

The participants also did not reach consensus on a recommendation that the products of agricultural biotechnology should be continued to be evaluated on a case–by–case basis using current regulations or methods. Some argued that the recommendation was unnecessary; others that we might not want to exclude the possibility of improving or changing the regulatory process.

COMMUNICATING WITH THE PUBLIC

This section of the report and recommendations is premised on a consensus agreement that the public has a stake in maintaining public institutions provided they are responsive to public needs. The decline in the credibility of scientists and public institutions should alert us to the fact that the public does not feel that its needs have been taken into consideration and that one of the reasons is the inability of the institutions and scientists to communicate with the public as equal partners in dialogue.

In the small group discussion the watchwords were: 1. *know your audience* and 2. *listen to what they have to say.* This is not as easy as it sounds because many (but not all) scientists have perceptions and biases that are quite different from the various perceptions and biases of public groups which makes it difficult for scientists to be good communicators. There is also the serious problem of lack of support for these activities in the reward structures of institutions and of an imbalance in funding going to high technology research versus research in policy and communications. These were all considered in the following set of recommendations, all of which were endorsed by all workshop participants.

—There is a body of knowledge about communications that scientists should use to improve the dialogue with the public. These include strategies like audience segmentation and use of focus groups. Ongoing survey research on scientist and consumer attitudes could be very helpful.

—*Regional research projects should be promoted and funded and the National Research Initiative should be encouraged to put more funding into its policy and marketing line item to promote public understanding of agricultural biotechnology.* Other agencies and entities such as foundations, other nonfederal agencies, industry and academia should be encouraged to promote such research and programs.

—*Interdisciplinary work between the biological and social sciences should be promoted and recognized as critical if serious progress in this area is expected.*

—*In all grant proposals the technical significance and relevance of research should be communicated in terms the general public (or anyone outside the particular discipline) can understand.* This is part of the ongoing discussion on the balancing of academic freedom versus public input into research priorities. At this point there are inadequate mechanisms for receiving input from those who do not have the knowledge and funds to lobby at the state and federal level. Advisory committees that have a broad representation of the public and heterogeneous interests should be constituted to work with colleges or departments directly.

—*Continuing education programs should be developed for scientists to teach them how to more effectively facilitate two–way communication between scientists and the general public.* Scientists need to learn how to recognize and understand the content and validity of a range of social, environmental and economic concepts that include the discussion of food safety issues by the public. They also need training in media relations and the communication process.

The workshop participants also recognized, as has been true in many other discussions of this type, that *the public, starting at the grammar school level, would be well served by educational programs on the social, moral, economic, political and scientific issues surrounding biotechnology.*

In order to accomplish any wide–ranging change in faculty behavior in these areas it will be necessary to re–envision the mission of the land–grant colleges to serve all their publics and recognize that the responsibility for this is shared by all institutions of higher education. This will change the weight given to public service or extension activities in promotion decisions and bring this area into better balance with research and teaching.

The workshop was quite remiss in failing to discuss in any detail the issue of labeling of products produced through biotechnology, and the contribution and relationship of labeling to communication with the public. We see this as an important topic for a future NABC meeting. *[Editor's note: the NABC 5 optional seminar will address the topic of labeling.]*

REGULATORY ISSUES

John E. Frydenlund
Deputy Assistant Secretary
Marketing and Inspection Services
U.S. Department of Agriculture

USDA Regulation of Animal Biotechnology

In this presentation I will share with you a U.S. Department of Agriculture (USDA) perspective on some of the regulatory issues associated with animal biotechnology. You are all aware of the Federal Coordinated Framework for biotechnology oversight that has moved products of biotechnology from the laboratory to the marketplace. Under this policy, federal agencies use their existing statutory authority to regulate the products of biotechnology.

The product reviews focus on the nature of the product and the risk, rather than the process used in its development. Federal agencies are required to ensure protection for public health and the environment from any potential harmful effects of these products. Favorable evaluations of the job the agencies are doing were published by the U.S. General Accounting Office and by the Office of Technology Assessment in 1988.

The "*Report on National Biotechnology Policy*," released by the Council on Competitiveness in the Office of the Vice President in 1991, reaffirmed the Administration's commitment to maintaining the U.S. lead in biotechnology research and product development over the long–term. The report provided an update on the Coordinated Framework and restated several principles for guiding federal regulatory policy:

—Federal oversight should focus on the characteristics and risks of the product, not the process, used in its development;
—Regulatory review should be designed to minimize burden while assuring protection of public health and welfare;
—Regulatory programs should be responsive to rapid advances in biotechnology.

The Council on Competitiveness' report also recommended publication of a document to help federal agencies make decisions, within the scope of authority afforded by statute, on how to regulate planned introductions of biotechnology products. The statement was published in February, 1992. It will ensure that federal oversight is used where it will do the most good—that is, where the risks are real.

We are using these principles to refine the management of our biotechnology programs at USDA. I would like to mention some of our management initiatives and then discuss the regulation of animal biotechnology.

USDA has both research and regulatory responsibilities for agricultural biotechnology and they are administered separately. The Assistant Secretary for Marketing and Inspection Services oversees the Department's biotechnology regulatory activities through a delegation of authority from the Secretary of Agriculture.

The counterpart for research is the Assistant Secretary for Science and Education. Together they co–chair the Committee on Biotechnology in Agriculture (CBA) which was established in 1986 as USDA's policy–making and coordinating body for biotechnology.

The members of the CBA are the administrators of all the USDA agencies that administer biotechnology programs—the Animal and Plant Health Inspection Service (APHIS), the Food Safety and Inspection Service (FSIS), Agricultural Research Service, the Cooperative State Research Service, the Economic Research Service and the Forest Service.

The CBA met recently (Spring, 1992) to review a strategy that we believe will improve the effectiveness of our biotechnology programs. Several of these initiatives are the result of a study of USDA management of critical issues, including biotechnology, that cut across the jurisdiction of individual agencies:

—We have solicited proposals for the biotechnology risk assessment research program stipulated in the 1990 Farm Bill. The purpose of the program is to strengthen the scientific basis of USDA's regulatory programs.

—We approved a public information plan on biotechnology for the Department. This is a priority program. We have been providing information for a long time on an agency–by–agency basis and the time has come to launch a Department–wide effort. We began the program this spring by co–sponsoring a joint U.S. and European Community (EC) meeting on biotechnology communication in Dublin, Ireland.

—USDA agencies are implementing the President's Biotechnology Research Initiative. This involves reporting and monitoring of the Department's $162.6 million 1992 research budget for biotechnology and reassessing our research priorities.

—We are committed to fostering trade through scientific meetings that will lead to international consensus on biosafety issues. USDA scientists are involved in the negotiations sponsored by the major international organizations including the EC and the Organization for Economic Cooperation and Development (OECD).

REGULATING PRODUCTS OF ANIMAL BIOTECHNOLOGY

Now I will turn to the activities of the two USDA regulatory agencies directly concerned with regulating the products of animal biotechnology—APHIS and FSIS. Both agencies have authority, in the broadest sense, for protecting animal health.

Animal and Plant Health Inspection Service (APHIS)

The Virus–Serum–Toxin Act of 1913, as amended, gives APHIS the authority to regulate all veterinary biological products imported into the U.S. or exported, or those biologics shipped or delivered for shipment interstate or intrastate.

Veterinary biological products are defined in the regulations [9 CFR 101.2(w)] as all viruses, serums, toxins and analogous products of natural or synthetic origin intended for use in the diagnosis, treatment or prevention of diseases of animals.

The licensing requirements [9 CFR Part 102] include tests to insure purity, safety, potency and efficacy. Pre–licensure evaluation of all veterinary biological products—regardless of the techniques used in their development—is performed at the National Veterinary Services Laboratory (NVSL) in Ames, Iowa. The NVSL is the only federal facility in the U.S. engaged in the evaluation of veterinary biologics and the diagnosis of domestic and foreign animal diseases.

It was nearly ten years ago that APHIS issued the first license for a veterinary biological product developed through biotechnological techniques. Since then, 48 product licenses have been granted for three broad categories of these products. The categories are based on the biological characteristics of the product and the kinds of safety issues it presents. Category one includes inactivated recombinant DNA–derived vaccines, bacterins, bacterin–toxoids and virus or bacterial subunits.

Hybridoma–derived monoclonal antibodies as well as genetically engineered antibodies are also included in this category. These nonviable, or killed, products pose no risk to the environment and present no new safety concerns. An example of a category one product is the *Escherichia coli* bacterin used to protect swine against Colibacillosis, a disease that has severe economic effects on swine producers.

The diagnostic test kits classified as category one products represent a significant breakthrough in animal disease diagnosis and treatment. Pseudorabies diagnostic test kits which can differentiate between reactions caused by wild type viral infections and immunization with recombinant vaccines, are used in APHIS' Pseudorabies Eradication Program.

Category two products contain live microorganisms that have been modified by the addition of marker genes or the deletion of genes that code for virulence. Special precautions are taken to ensure that the addition or deletion of genetic information does not confer virulence, pathogenicity or survival advantages to these organisms that are greater than those found in the parent or wild type forms. All the licensed category two products are for use against pseudorabies in swine and involve gene deletions or additions.

Category three includes products containing live expression vectors carrying recombinant DNA–derived sequences that code for immunizing antigens or other immune stimulants. The transmission characteristics of such products must be carefully assessed before field studies are undertaken.

While no licenses have been granted in category three, one product is currently being field tested after a thorough evaluation of safety data. The product is a live recombinant DNA–derived vaccine–vectored rabies vaccine intended for oral use in raccoons in the wild. The incidence of rabies has increased dramatically in the Mid–Atlantic States and public health officials have been enthusiastic about the potential of immunizing animals in the wild with the recombinant vaccine contained in food bait. Field tests have been conducted on Parramore Island in Virginia and in Sullivan County, Pennsylvania.

Additional tests have begun in a three–county area in New Jersey. Since December, 1989, there have been 1700 cases of rabies in New Jersey and 90 percent of the cases have been in raccoons. The disease is moving from North to South in the State and New Jersey public health authorities hope to establish a rabies–free zone by concentrating the bait drops in Cape May County. There will be extensive monitoring before and after another bait drop in Fall, 1992.

The states must approve field tests of all experimental biologics so we have worked closely with state officials to provide the public with information on the rabies vaccine field trials. APHIS scientists attended state–sponsored public meetings to answer technical and scientific questions and an APHIS spokesperson was interviewed in the Spring, 1992, on Cable News Network (CNN) about the New Jersey tests. The vaccine trials were also featured in a new publication on biotechnology distributed in March, 1992, to junior high school students throughout Pennsylvania.

Future generations of vaccines will combine the genetic information to immunize against several diseases into one virus or bacteria. We have now received an application for a genetically engineered category three vaccine with antigens against two disease agents in the same microorganism. This will improve the consistency of production from lot to lot of product, by eliminating variation in antigen content when different components are mixed together.

For the category two licenses and the field trials of the category three rabies vaccine, APHIS prepared a complete environmental assessment of the proposed action in compliance with the provisions of the National Environmental Policy Act. The environmental assessments provide the public with a discussion of scientific data on safety and a thorough analysis of environmental impacts.

When any project has implications for public health, expert panels consisting of representatives from federal agencies, academic institutions and professional societies are convened to review data. Additional review can also be requested from the National Vaccine Program in the Department of Health and Human Services which was done for the rabies vaccine.

Early generations of veterinary biological products developed through biotechnology have proved effective in disease prevention and diagnosis. Succeeding generations will be even more effective and we look increasingly to global markets for these products.

The growth of international ownership of the biologics production industry and the advent of the European Community (EC) have increased the immediacy of the drive for internationally recognized standards and consistency in testing procedures. We have worked with a number of international groups to further this process. In one of these efforts, U.S. and EC representatives met in France in January, 1992, to work toward the standardization of production practices. The discussions continued during that summer.

There is pressure on the U.S. to increase the potential for EC–produced biologics to enter U.S. markets. Many of these products are prevented entry because of the presence of foot and mouth disease and bovine spongiform encephalopathy in several EC countries. We are working to resolve these and a number of other issues.

Food Safety and Inspection Service (FSIS)

Turning now to the role of the Food Safety and Inspection Service in regulating animal biotechnology, this will be brief because Dr. Cross has already done such an able job on this subject (see page 121). The summary statement from last year's NABC workshop on transgenic animals is a good place to begin and I quote:

> In general, workshop participants concluded that public issues associated with transgenics were not urgent, primarily because applications of transgenic technologies as they affect agriculture and the food supply seem remote at the present time. (Murray et al., 1991, p. 43)

We think that if we do our work well enough, public issues associated with transgenics will *not* become urgent. This means that we must continue to maintain an open dialogue on potential issues before they develop. We talk to our critics, as well as to the regulated public, and we join with other federal agencies in sponsoring workshops and discussion groups.

These statutes from which FSIS gets its authority for regulating the animal products of classical breeding and new technologies are the Federal Meat Inspection Act and the Poultry Products Inspection Act.

These statutes require that FSIS inspect cattle, sheep, swine, goats, equine, poultry and food products prepared from them intended for use in human food to assure that they are wholesome, not adulterated and properly labeled, marked and packaged. The FSIS policy statement in the 1986 Coordinated Framework established the applicability of the experimental animal regulations for the use of genetic engineering techniques in food animals [9 CFR 309.17 and 381.75].

This policy was reaffirmed in a *Federal Register* notice published in December, 1991. The notice pointed out that only a small proportion of the animals in gene transfer experiments contain the intended gene [56 FR 67054-67055]. Before animals that do not contain the experimental transgene may be presented for slaughter, data must be submitted to FSIS demonstrating that the transgene is not present and that the animals are therefore "not adulterated." Written approval for slaughter of an animal determined to be nontransgenic is granted by the FSIS Deputy Administrator for Inspection Operations and the animals are subject to the same inspection procedures as conventionally bred animals. Animals have been approved for slaughter under these provisions. We expect a number of applications to be considered in the near future. We know that the cost of maintaining these animals is prohibitively high.

A document is being prepared which pertains to the food safety evaluation of transgenic animals being considered for slaughter. In making an evaluation, FSIS may consult with other agencies, including the Food and Drug Administration (FDA), the Environmental Protection Agency (EPA) and APHIS. Interaction among federal agencies to consider these issues has been taking place for several years through the meetings of the Food Animal Biotechnology Information Exchange Group. Through these discussions we are working to anticipate both scientific and consumer issues, and to avoid delays due to agency concerns about jurisdiction.

We know that research using molecular methods will bring revolutionary advances in animal science. The realization of much of this promise is in the future for the development of animals bred for special qualities such as disease resistance. However, this research has already brought us quantum increases in our knowledge of gene function. There have been many notable breakthroughs. Transgenic animals have proven very useful as models for the development and treatment of a variety of human diseases. The use of large animals as bioreactors has resulted in the development of sheep that secrete such substances as Clotting Factor 9 in their milk, and transgenic pigs that carry quantities of human hemoglobin in their blood.

There are any number of applications of biotechnology to animal science and production agriculture that are scientifically successful including the production and use of bovine and porcine growth hormones. One application, in particular, must be singled out because it relates directly to the FSIS responsibilities for food safety. The availability of DNA probes to test for the presence of bacterial pathogens in meat and meat products has cut the detection time by one–half. These molecular methods are also highly sensitive and cost–effective.

The use of existing statutes and procedures to evaluate the products of animal biotechnology has allowed us to anticipate the testing and marketing of new products. We have ensured our capability for dealing with the new techniques by hiring specialists and through training programs for our staff scientists. We urge researchers to meet with us to discuss any questions they may have about regulations for testing and product development.

CONCLUSION

In closing, I would like to emphasize the point Dr. Cross made about the importance of consumer interests and perceptions (see page 125). USDA agencies, including FSIS and APHIS, work closely with consumer–interest groups to inform the public about our oversight policies and programs for biotechnology products and to discuss any safety concerns associated with production and marketing.

The potential of biotechnology for improving animal health and the quality of meat and meat products is immense. We believe that risk–based regulatory programs will help realize the benefits of the technology for both U.S. consumers and producers.

REFERENCES

Murray, J., P.B. Thompson and R. Piggott. 1991. Transgenic Animals. In *NABC Report 3, Agricultural Biotechnology at the Crossroads: Biological, Social and Institutional Concerns.* J. Fessenden MacDonald, ed. National Agricultural Biotechnology Council. Ithaca, NY p. 43.

Martin Terry
Vice President for Scientific Activities
Animal Health Institute

Animal Pharmaceuticals

UNDEFINED TURF

The Food and Drug Administration (FDA) regulates animal drugs under the Food, Drug and Cosmetic Act, while the U.S. Department of Agriculture Animal and Plant Health Inspection Service (USDA/APHIS) regulates animal biologicals under the Virus–Serum–Toxin Act. But the question of which agency has jurisdiction to regulate a given animal health product is not always a totally obvious one. The regulatory definition of a biological is in need of updating to clarify the status of some of the compounds being developed through biotechnology. Particularly difficult to classify are those compounds which occur endogenously, modulate the immune response and have pharmacological properties. A classic example of this sort of regulatory ambiguity can be seen in the handling of the interferons which are regulated as *biologicals* (by FDA's Center for Biologics Evaluation and Research) *for human use*, but are regulated as *drugs* (by FDA's Center for Veterinary Medicine) *when labeled for use in animals.* This disparity has more to do with interagency politics than it does with the pharmacological/immunomodulatory effects of interferons. The Animal Health Institute (AHI) is currently working on a proposal to amend the regulatory definition of a biological, with the object of providing a more adequate taxonomy of drugs vs biologicals—which would *ipso facto* determine which agency should have jurisdiction to regulate a given substance or product.

PRODUCT VS PROCESS AND THE GLASS FOURTH HURDLE

A cornerstone of the final "Scope" policy statement of the Office of Science and Technology Policy as published in the *Federal Register*, February 27,1992, is the notion that regulatory oversight is appropriately applied in direct proportion to the *risk* associated with a given *product per se,* independent of the *technology* employed in the manufacturing *process.* Interestingly, neither objective product risk assessment nor concern with the nature of the manufacturing process has occupied center stage in the controversy surrounding bovine somatotropin (BST), the first high–profile product of biotechnology to be developed as an animal drug.

Since FDA finished its food safety evaluation of BST in 1986 and pronounced that there were no human food safety issues arising from the use of BST in lactating dairy cattle (Juskevich and Guyer, 1990), the public debate has focused on possible economic and social effects of the anticipated widespread adoption of the use of BST by the dairy industry. Questions have been raised and projections made as to the magnitude of the effects of widespread BST use on volume of milk production, milk prices, dairy herd size and the continued viability of marginal, inefficient dairy operations.

FDA cannot legally take such socioeconomic considerations into account in the premarket drug approval process; animal drugs must be evaluated on the basis of the objective criteria of safety and efficacy. However, FDA does not operate in a political vacuum and in a situation where heated political debates on the socioeconomic aspects of a new animal drug run concurrently with the regulatory evaluation of the drug, it is hard to believe that the agency would not be affected to some degree in its deliberations on the drug. At the very least, the political heat radiating from the socioeconomic issues can be seen to make the agency even more cautious than usual in its evaluation of the safety and efficacy data on the drug—which would logically result in a delay in the approval process.

Moreover, while the effects of socioeconomic criteria on the regulatory process may be subtle and unofficial at the federal level, they can be blatant and most official at the state level, as evidenced by the current legislative moratorium on the use of BST in Maine.

DOORS VS WINDOWS

Another issue which is by no means unique to biotechnology products, but which, as a matter of historical fact, emerged as the subject of public controversy in the course of the ongoing BST debate, is that of *regulatory transparency.* In 1986 in the UK, headlines appeared about "secret trials" being conducted with BST on undisclosed farms, with innuendoes of collusion between the animal health companies and the British Ministry of Agriculture. What, in fact, was happening was that animal health companies were field–testing a product for efficacy—the human food safety of the product having already been established to the satisfaction of the regulatory authorities—in accordance with the pertinent regulations, in exactly the same way as hundreds of other animal health products had been tested previously. The brouhaha arose, not because BST was receiving any favored treatment by the British government—that was clearly not the case—but rather because the public was totally unaware of the regulations and legal procedures routinely used in the testing and approval of animal drugs—until the BST critics sought to portray those procedures as some sort of conspiracy against the public.

At the heart of the transparency issue is a conflict between the public's "right–to–know" and a drug sponsor's legal right to confidential treatment of proprietary information on a product including the details of the tests conducted to demonstrate the safety and efficacy of the product in the regulatory approval process.

In the U.S., the Food, Drug and Cosmetic Act provides that data which are submitted on a new animal drug by the drug sponsor will be evaluated on a confidential basis by FDA. The first regulatory hurdle which must be cleared in the drug approval process is that of establishing human food safety. Only when a new animal drug has been sufficiently evaluated and found to be safe from the point of view of humans consuming food produced by animals treated with the drug, may FDA approve an Investigational New Animal Drug Application (INADA) which authorizes the drug sponsor to conduct tests to demonstrate the efficacy and target animal safety of the drug. In the FDA–monitored field trials conducted under an INADA, the issue of the safety of the food derived from the test animals has already been resolved by FDA. Thus, in terms of human food safety the drug is no longer an "experimental drug." At the INADA stage, a drug is actually "experimental" only in regard to its efficacy and its safety *to the target animal* at dosages intended for commercial use. The food produced by the animals involved in such testing is as safe as any other food produced with fully approved new animal drugs. Claims that the public is being put in jeopardy through exposure to food produced with an "experimental (INADA) drug" are ill–founded, and usually mischievous.

Yet there is a point of view which says that even if FDA says a product or technology is safe, the public has a right to know whether the food in commercial channels was produced with that product or technology. Here the question of *labeling* rears its head and labeling is a highly controversial issue.

In principle, no one should object to providing the consumer with as much objective, nonproprietary information about food products as the consumer has patience to read. However, in the reality of commercial food production and marketing there are some difficulties involved in routinely providing certain types of information on the label of a food product. Let us consider, for example, the very topical notion of positively labeling a food product with something like: "Produced with xenophobein, a hormone derived from recombinant DNA technology." Let us say that this imaginary protein called "xenophobein" has been found by FDA to pass all the rigorous regulatory tests for human food safety and that food produced with xenophobein is analytically identical to food produced with traditional technologies. In such a case, what are the consequences of putting the above–quoted information on the label of food produced with xenophobein? Four come immediately to mind:

1. Such labeling contributes *nothing* to the consumer's knowledge in terms of safety or nutritional information, given that there is no objective difference between food produced with xenophobein and food produced without it. So, in a scientific sense, such labeling is gratuitous and of *no* consequence.

2. To insist on detailed labeling as to the technology by which a food is produced when the food itself has been found by FDA to be safe, is essentially

to discount FDA's safety evaluation process. If we accede to demands that consumers should be given the opportunity to make their own safety assessment (presumably on the basis of data purveyed by such prestigious scientific journals as *The Wall Street Journal*), the logical conclusion is that product labeling should be sufficient to allow consumers to protect themselves through the exercise of "informed" choice. But in that case FDA's evaluation would be, at best, redundant and, at worst, in disagreement with the consumer's personal evaluation. We could just as well dispense with the services of FDA and revert to a system of exhaustive labeling and *caveat emptor*.

It seems to me that, for all FDA's imperfections—and AHI is traditionally one of FDA's most vocal critics—we as a society are better off with a government–run agency making regulatory decisions on the basis of expert scientific evaluation, than we would be in a system of "every man [sic] his own regulator," where some would demand that an encyclopedia of product information and manufacturing data be attached to each can of pork 'n beans and the role of the agency would be reduced to that of an editor of encyclopedias.

3. As my colleagues in the European Commission learned when they proposed positive labeling of beef produced with hormones as a solution to the European hormone debate in the mid1980s, the use of emotive terms like "hormone" in labeling is likely to scare, rather than objectively inform, consumers. (That, of course, is exactly the effect desired by many who advocate such labeling, as their interest is not in accurately informing consumers, but rather in politically motivating consumers through the manipulative use of "hot" language which serves to obfuscate rather than to educate.) Even with the best of intentions, what is intended as a neutral statement of fact on a label can be all too easily misinterpreted as a warning.

4. Labels generally are, of necessity, minor masterpieces of succinctness. Space on a label is available only at a cost and any statement that did not convey concrete information as to the safety or nutritional value of the contents of a food package would carry uncompensated added costs which would increase the cost of the product to the consumer without providing a benefit.

I would emphasize that the above considerations apply only to *positive* labeling, e.g., "This product was produced with xenophobein." There are no such objections to *negative* labeling, e.g., "This product was produced without xenophobein." Though the fact remains that the product produced *without* xenophobein is identical to that produced with xenophobein—and therefore the negative labeling is scientifically as meaningless as positive labeling would be—the option of negative labeling allows consumers to exercise choice in the process by which their food is produced.

Negative labeling has the advantage of being equitable to all parties involved in food production and consumption. If there is sufficient demand for a product produced by an "alternative" (which generally means low–tech, high–cost) technology—i.e., if producers find that it is commercially viable to exploit a niche market based on a perceived consumer preference for food produced without the use of a "mainstream" (generally high–tech, low–cost) technology—then by all means "let a thousand flowers bloom" in the marketplace. Let consumers decide with their checkbooks which products best satisfy individual preferences—whether those preferences be based on cost, safety, nutritional quality, aesthetics, ideology, or a combination of factors. Let producers decide which markets they want to cater to, matching production technologies with consumer preferences—as determined by the extent to which consumers are in fact willing to pay premium prices for products produced with the less efficient technologies.

COMMERCIAL PROMOTION VS POLITICAL SELF–DEFENSE

Of current concern in the regulation of animal pharmaceuticals—and of particular relevance to those derived from biotechnology—is the issue of pre–approval "promotion" or defense of a product by the manufacturer while the product is still under evaluation by FDA. The agency has defined "promotion" so broadly as to impose very narrow limitations on the information that can legally be conveyed to the public by a manufacturer about a product in the pre–approval phase. The AHI has taken a quite different position on what kinds of activities constitute commercial promotion—as opposed to the pre–approval defense of a product in response to political attacks intended to prevent its approval.

The AHI view is that the severe limitations on the dissemination of information which FDA has sought to impose on AHI and its members are not dictated by the relevant regulations, are inconsistent with Administration policies to remove impediments to the development of new technologies and are seriously at odds with AHI's and its members' constitutionally protected rights to protect their property interests fully in the political arena. Negotiations are underway to attempt to resolve this dispute which has far–reaching implications for the application of biotechnology in animal agriculture.

IMPEDIMENTS TO DRUG APPROVAL VS THE THREAT OF SNAKE OIL

The last issue I would like to touch on, namely, the need to revisit the statutory efficacy standard for animal drugs, is in a sense a by–product of the current debate on extra–label drug use. It is not unique to biotechnology products, but it has coincidentally arisen as a major regulatory issue at the time when the first biotechnologically produced animal drugs are in the latter stages of the FDA approval process—which is to say, at the time when these biotechnology products are undergoing efficacy testing.

The efficacy standard for animal drugs, as set forth in the Food, Drug and Cosmetic Act, is relatively simple and straightforward. In ordinary language, FDA must require "substantial evidence" that a product is effective for the purpose for which it is intended, i.e., an animal drug cannot be approved by FDA until it has been shown not to be "snake oil." The "substantial evidence" required is ordinarily in the form of "adequate and well–controlled studies."

Unfortunately, this reasonable, bare–bones efficacy standard as set forth in the statute has undergone a sea–change of agency interpretation over the last few decades. Regulatory barnacles and mineral accretions have built up in the form of ever–more–complex policies for efficacy testing requirements. The problem has now reached such proportions that it often costs more for an animal drug sponsor to conduct the efficacy studies required for FDA approval than it does to conduct the *safety* studies which have traditionally accounted for the major portion of the cost of product development.

Particularly onerous to sponsors seeking regulatory approval of new animal drugs is FDA's current policy to require "optimal dose" titration studies, in which a number of doses (the majority of them irrelevant to clinical reality) are tested for efficacy, in order to determine the lowest dose which is adequately effective. Clearly this requirement adds greatly, and unnecessarily, to the cost and time required to get a drug approved. But it also has the additional long–term disadvantage of freezing the label dose at a level which may itself be clinically irrelevant by the time the drug has been used in the field for a few years. (For a good example of the latter problem, consider penicillin which is now universally acknowledged to be virtually useless in veterinary medicine at levels less than three times the once "optimal" dose which is still the only dose on the product label.)

To address these problems, AHI and the American Veterinary Medical Association (AVMA) have filed a Citizen Petition proposing that FDA approve dosage ranges and eliminate the requirement for optimal dose titration in efficacy testing. The proposal applies only to new animal drugs which have adequate safety data packages and which would be restricted to use by or on the order of a licensed veterinarian. The statutory deadline for FDA to respond to the Citizen Petition passed silently several weeks ago. In light of the recently launched AVMA legislative initiative to legalize extra–label drug use by veterinarians, this silence on FDA's part could be pivotal. If FDA is perceived as being unable to respond positively to proposals submitted in a regulatory mode to streamline current policies on efficacy testing, that would surely be interpreted by some in the industry as an indication that legislation is the only available remedy.

Regardless of whether it is eventually achieved through regulatory channels or through Congress, streamlining the efficacy testing requirements for animal drugs has clear advantages. It would free up agency resources to devote to the crucial process of safety evaluation of new animal drugs. It would remove a major economic disincentive that currently discourages drug sponsors from seeking broader label indications for new drugs. And it would have a positive impact on the rate of approval of new animal drugs (including biotechnology products) which would mitigate the crisis in drug availability which currently besets animal agriculture.

REFERENCES

Juskevich, J.C. and C.G. Guyer. 1990. Bovine growth hormone: Human food safety evaluation. *Science.* 249: 875.

Office of Science and Technology Policy. 1992. Exercise of federal oversight within scope of statutory authority: Planned introductions of biotechnology products into the environment. *Federal Register.* 57: 6753. February 27.

Margaret Mellon
Director, National Biotechnology Policy Center
National Wildlife Federation

The Regulation of Genetically Engineered Animals: Going From Bad to Worse

Genetically engineered animals raise a number of issues that might potentially call for regulation. Animals may pose human health risks if used as food, environmental risks if they escape from confinement, or animal health or food quality issues where they are genetically adapted to produce valuable drugs.

Since genetic engineering is capable of producing so many different kinds of animals for so many purposes, a thorough analysis of the adequacy of current regulation is well beyond the scope of this paper. Here I will confine myself to environmental risks and summarize the current regulatory climate. What I have to say is genuinely discouraging. Not only are animals not being adequately regulated now, but, under the Bush Administration, there is little chance that they ever will be.

GENETICALLY ENGINEERED ANIMALS: ANYTHING BUT NATURAL

We must begin with the basic point. Genetic engineering is a radically new technology, and anything but natural. Modern gene transfer technologies permit artificial gene transfers across species, family and even kingdom lines. Genes from cows can be put in fish; genes from butterflies into tomatoes; genes from moths into potatoes. With genetic engineering the number of potential new combinations is almost limitless. While we now have considerable experience with genetically engineered organisms, most of it involves microorganisms used under laboratory conditions. We have much less experience with environmental release of engineered organisms. In the case of released animals, we have almost no experience.

While the process of genetic engineering is not inherently dangerous and should not by itself lead to prohibitions, it does have the potential to modify organisms' traits and behavior in ways that are not well understood and not easy to predict. Particularly where the animals will be released—accidentally or deliberately—into the environment, the process of engineering creates sufficient uncertainty that it warrants a red flag of caution.

ENVIRONMENTAL RISKS

Organisms with new combinations of genes can exhibit new combinations of traits, and new combinations of traits may enable new behaviors in the environment. How organisms with new traits will fare in the environment is difficult to predict. The factors influencing success and competition are many and complex. But new trait combinations have the potential to improve an organism's chances of success in the environment. If this happens, an organism can displace existing organisms or otherwise disturb existing ecosystems.

Consider fish. In contrast to some domesticated animals such as cows or sheep, which are unlikely to survive in the wild without human assistance, fish are wild animals well adapted to their environment. Even small modifications by genetic engineering could equip them to survive in a broader range of habitats.

A good example of a one gene change that could have dramatic impact on fish survival involves the so–called "antifreeze" gene. Such genes, available from flounder, code for proteins which can keep fish blood from freezing in arctic waters. So far, antifreeze genes have been transferred into several warm–water fish, including carp. With the antifreeze protein in their blood, the warm–water carp can survive in cold waters where they might displace native cold–water fish or in other ways disturb the aquatic ecosystem.

Another example involves genes for growth hormones. Auburn University scientists have recently transferred growth hormones from other fish and mammals into carp and catfish. Like cold tolerance, fast growth can enable a fish to displace other species and disrupt food chains. In both cases, the new gene can move into any fish that can breed with the engineered fish.

Generally, the same concerns apply to the release of any genetically engineered animal into the environment. Whether an insect, a snail, a mouse or a cat, animals with modified growth rates or temperature tolerances or a multitude of other new traits pose the risk of disrupting ecosystems in harmful ways. Since genetic engineering has the ability ultimately to transfer an unlimited number of new traits into animals, its risks are likely to be greater than those posed by traditional breeding.

Moreover, in addition to the desired modifications, some gene transfers may have effects the genetic engineers did not want, and could not predict. A sad illustration is afforded by the so–called Beltsville pig. In this case, researchers at U.S. Department of Agriculture's (USDA) Beltsville laboratory succeeded in transferring human growth hormone genes to pigs in hopes of producing leaner meat. Instead, the pigs have proven to be crippled, cross–eyed and immune–compromised. Other, less obvious, secondary effects may occur with other gene transfers. Some of these may affect behavior and impact on release.

Finally, as shown by the unexpected effect of chloroflourocarbons (CFCs) on the earth's atmosphere, our ability to predict the impacts of technology is limited. It is possible that genetic engineering, too, may pose novel risks that we have yet to appreciate.

THE FEDERAL BIOTECHNOLOGY REGULATORY FRAMEWORK

The American Fisheries Society (AFS) has recently issued a position statement highlighting the risks of the release of genetically engineered fish and concluding that such releases ought to be overseen by government (Kapuchinski and Hallerman, 1990).

Unfortunately, the American Fisheries Society also noted that the Federal Coordinated Framework does not require the necessary oversight.[1] And the Fisheries Society is correct. No comprehensive federal authority exists under which the releases of fish will be reviewed. In fact, little authority exists to control the environmental impacts of any genetically engineered animals—be they fish, fowl or insect. *Right now, anyone who wished to genetically engineer and release a frog—or for that matter, mink, dog or rat—into the environment is generally free to do so without fear of federal repercussions.*

This information may surprise some readers who perhaps believe that the federal government has a comprehensive framework in place. In fact, the framework that does exist ignored, from its inception, the environmental impacts of animals. I will say a few words about its inadequacies below, but unfortunately, the framework is rapidly becoming a moot point. The Bush Administration engaged in an effort not simply to stall or weaken implementation of the framework, but to dismantle it entirely.

APPLICABILITY TO ANIMALS

Before I turn to the 1992 efforts of the Bush Administration, let me briefly touch on the history and components of the federal framework and its relevance to the control of environmental risks posed by animals.

In 1983, the Reagan Administration orchestrated a multiagency effort to develop a policy to regulate biotechnology and its products. Operating out of the Office of Science and Technology Policy, Administration officials gathered together representatives of USDA, the Food and Drug Administration (FDA), Environmental Protection Agency (EPA) and other agencies to evaluate the statutes they administered for their applicability to biotechnology. In 1984 and 1986, the Agencies published statements outlining how their statutes would be applied to products expected from the new technology. Those statements constitute the core of Federal Framework for biotechnology regulation.

Perhaps because so few engineered animals were under development in the 1980s, the Framework said little about the environmental risks posed by animals.[2] Although a few of the statutes that make up the framework could potentially be applied to releases of animals, no attempt has been made to do so.

---

1 USDA officials have indicated that, instead of FIFRA, they intend to use the weaker authority of the Plant Pest Act to regulate genetically engineered pest control agents. See Payne, 1992.

2 Almost all the discussion of animals related to the health effects of genetically altered animals used as food.

One of these statutes, the Plant Pest Act, for example, could apply to animals that meet its definition as plants pests. Release of such organisms, which might include genetically engineered insects, slugs or nematodes, could be covered by the existing Plant Pest Act program. The statutory definition of a plant pest, however, is severely restricted—covering only invertebrates. Vertebrate animals, like fish or frogs, are completely excluded from the coverage of the statute.

The Federal Insecticide, Fungicide and Rodenticide Act (FIFRA) is another statute under which animals could potentially be regulated. Insects that prey on pest organisms, for example, are considered pesticides under FIFRA. Engineered animals developed for this purpose would be subject to regulation as pesticides under FIFRA. Currently, however, EPA exempts invertebrate animal pesticides from regulation under FIFRA on the grounds that such animals are adequately overseen by the USDA. USDA, in turn, has made little effort to implement its authorities to regulate animals (Payne, 1992).

It should be also noted that EPA's interpretation of the jurisdiction of the Toxic Substances Control Act (TSCA) is broad enough to cover all living organisms, including animals, but that as a matter of policy the Agency has restricted its TSCA program to microorganisms.

Finally, the National Institutes of Health (NIH) Guidelines require NIH approval for the release of genetically engineered animals used in research. The guidelines, however, cover only government–funded research. Private enterprises, for example, commercial fish farms or pet breeders, are not covered by the guidelines.

In summary, except for animals used in federally funded research, developed for pest control or invertebrate animals classified as plant pests, the release of genetically engineered animals is not, and does not have the potential of being, regulated under the Federal Framework for the Regulation of Biotechnology. *From anemones to zebras, most animals can be engineered by anyone, for any purpose, and released at will.*

THE COUNCIL ON COMPETITIVENESS

In retrospect, the evolution of the current (1992) biotechnology policy has been a steady downhill slide. The Biotechnology Regulatory Framework developed by the Reagan Administration promised at least four new rules or guidelines: two implementing USDA authorities[3] and two implementing EPA authorities. Six years have now passed. Only one of the promised regulations has been promulgated—USDA's regulations under the Plant Pest Act. Neither the EPA regulations under the TSCA or under the Pesticide Act, nor the USDA guidelines governing agricultural research have seen the light of day.

[3] One of the promised guidelines would have implemented the USDA research authorities to establish a set of guidelines governing the release of genetically engineered animals in research. These guidelines would have covered an important set of activities involving genetically engineered animals.

The Bush Administration (working through the Council on Competitiveness in the Office of the Vice President) has gone beyond blocking implementation of the framework. It is now trying to dismantle programs—specifically the Plant Pest Act program—already in place.

The withdrawal from the arena of biotechnology policy was not instigated by the relevant Agencies. To the contrary, both the USDA and EPA, implementing the Federal Framework, have sent successive versions of proposed rules and guidelines to the White House for approval. Their efforts have been blocked by the group currently responsible for this policy—the Council on Competitiveness .

With regard to animals, an adequate regulatory framework would require both new legislation and implementation of existing laws in ways not contemplated in the 1986 framework. Since the Council on Competitiveness will not allow even the implementation of the laws promised in the 1986 document, there is no hope for the new initiatives needed for animals.

SUMMARY

As of this NABC meeting (May, 1992), the regulation of genetically engineered animals is hopelessly inadequate, with little hope for improvement. As long as the Council on Competitiveness sets policy, existing statutes are unlikely to be implemented to regulate genetically engineered animals and no new legislation will be sought to provide the new authority needed.

From an environmental standpoint, the current situation means that the risks posed by engineered animals to the environment—whether from accidental or deliberate release—will go unassessed and uncontrolled. Moreover, without regulation there will be few opportunities for the public to know what is coming or to participate in decisions about the technology. The bottom line is that the new policy leaves it up to industry and scientists to decide what kind of animals to make and when and how they should be released. The rest of us must simply hope that their choices will not lead to environmental degradation and disaster.

This policy of secrecy and exclusion of the public is a recipe for disaster–both for the environment and for the biotechnology industry.

REFERENCES

Kapuchinski, A.R. and E.M. Hallerman. 1990. AFS Position Statement: Transgenic Fishes. *Fisheries.* 15:2-4 August.

Payne, J.H. 1992. Regulation by USDA/APHIS of Biological Control Organisms That Have Been Genetically Engineered. In *Regulations and Guidelines: Critical Issues in Biological Control.* R. Charudattan and H. W. Browning, eds. Proceedings of a USDA/CSRS National Workshop, June 10-12, Vienna, VA.

Bennie Osburn
Associate Dean, Research and Graduate Education Programs
School of Veterinary Medicine, University of California, Davis
Robert B. Nicholas
Partner, McDermott, Will & Emery
Washington, DC

Workshop Report

The charge to the workshop participants was to identify and examine issues arising in regulatory treatment of animal biotechnology. This charge was addressed in an open forum in which free ranging discussion among individuals with different perspectives was strongly encouraged. The two introductory talks by Martin Terry, Vice President for Scientific Activities, Animal Health Institute and Margaret Mellon, Director, National Biotechnology Policy Center, National Wildlife Federation, gave rise to an initial discussion. The approach taken by the group was to list relevant issues, group those issues into three basic categories, discuss the issues category by category and develop shared issue statements or recommendations. The issues raised by various members of the groups are listed below according to category. While consensus was not sought nor achieved on the specific issues listed, these issues were deemed worthy of consideration by one, some or many of the members of the group and, as such, help to illustrate the range of concerns in this arena. Common themes of agreement did emerge and these were captured in the form of four issue statements or recommendations that represent points of consensus and, as such, they highlight important underlying concerns in this arena. The three basic categories discussed are as follows:

THE REGULATORY PROCESS *(how the process works)*
This section deals with: 1. How the system works; 2. The issues in formulating and implementing regulations; and 3. Where there are gaps in the system that are of potential safety and/or environmental risks. The following issues and gaps have been identified:

Research Stage

The National Institute of Health has not adopted Appendix Q which contains guidelines for contained research on transgenic animals. These guidelines would be helpful for Institutional Biosafety Committees and others. There are no mandated guidelines/regulations for industrial research of animal biotechnology.

Clinical Testing of Drugs
No obvious shortcomings were identified.

Field testing

—There are no regulations for release of fish, wildlife, insects or pets; for micro–organisms in livestock feeds; or for zoonotic pathogens of animals and humans;

—Implementation of "Guidelines for Research Involving Planned Introduction into the Environment of Genetically Modified Organisms" developed by the Agricultural Biotechnology Research Advisory Committee (ABRAC) should govern agricultural research;

—Absence of mechanism to deregulate similar genetically modified organisms that have been proven to be safe based upon previous case studies;

—The inability to gain access to some information on health and safety of products because of "confidential business information" designation.

Large Scale/Commercial Release
There are no oversight mechanisms, guidelines or regulations for large scale commercial release:

—Of products of animal biotechnology;

—Of second to nth generations of transgenic animals;

—Impact of the production system for the environment (e.g., genetically modified organisms replacing indigenous populations).

Regulatory Assessments (current regulations that relate to animal biotechnology)

Animal safety: the current regulatory system for drugs and therapeutics appears to satisfactorily cover animals that receive or are altered by biotechnology.

Food safety: foods of animal origin are regulated by USDA's Food Safety Inspection Service and the FDA's Center for Veterinary Medicine for safety, quality and efficacy. The gaps that are currently not covered include disposition of transgenic animals such as: Should they be reviewed case by case? Should transgenics receive prolonged testing before slaughter? Should fish, seafood and wildlife not be covered?

Efficacy: the efficacy of drugs is regulated by the FDA. Should transgenic animals used as pharmacoreactors receive special attention?

Market Place
Should consumers have the ability to make choices by knowing when they are purchasing products resulting from animal biotechnology?

ANIMAL BIOTECHNOLOGY REGULATIONS AND PUBLIC INVOLVEMENT

(who is involved in regulations)

The Public's Role in Debate

There should be representation of broad interest. The access to information and participation in debate should be improved for interested persons.

A clear definition of process (where input and questions, etc. can be integrated) should be known so that those wishing to participate could do so. Channels for participation may vary across agencies.

Possible mechanisms for improved access include:

1. Legislation regarding public participation in regulating decisions across the board;
2. Publication beyond the *Federal Register*;
3. Improved representation in decision–making processes;
4. Open forums;
5. Research on opening up scientific decision–making process; and
6. Rebuilding public trust and regulatory transparency.

Other Issues To Be Considered

—Role of states and industry in debate;

—Public education and who has responsibilities for keeping the public informed;

—Communication and knowledge can lead to choices by the public; and

—Labeling products developed through biotechnology.

TECHNICALLY BASED REGULATIONS VS SOCIAL/ETHICAL/ECONOMIC IMPACT CONSIDERATIONS *(What is the basis for regulations? Why are decisions made?)*

Regulatory Impact

The goals of regulations include:

1. Safety of the public and environment;
2. Safety and efficaciousness of food;
3. Quality assurance of products; and
4. The safety and welfare of animals.

Regulations can impact not only in the U.S., but also on international trade as well as trade relationships with third world countries. Patenting, however, impacts as a socioeconomic factor.

Information and Consumer Choice

Consideration was given by the group to the level of information available for consumer choice:

—Is there a need for labeling which would provide the public with a way to reflect their individual values?

—Should labeling be voluntary or mandatory?

—What information should be made available?

—What are the criteria for labeling?

—Can labeling be used as an education device? Concerns voiced by participants included complexity of labels, definitions, etc.

The group acknowledged early on that many groups of people (e.g., minorities, farmers, industry) were not well represented in its deliberations; the workshop's report stems from a lower diversity of backgrounds than might be wished. Nonetheless, a wide diversity of positions relative to animal biotechnology regulations were represented. It is the hope of the group that its recommendations may contribute to positive actions and that its listing of issues may stimulate further debate in many other forums.

RECOMMENDATIONS

1. *The regulatory gaps delineated deserve serious investigation. NABC may wish to establish a committee or other mechanism to assist this investigation.*

2. *A more acceptable policy–making process for rules of broad applicability would be clearly understood or known (not ad hoc), transparent and participatory. The group viewed the process leading to the recent FDA food safety decision as falling short of the goals for an acceptable process.*

3. *Social, economic and ethical questions need to be explored. What role do/should these issues have in research, development and approval processes for commercial use of new products? When should these factors be considered, relative to, but not necessarily as a part of the regulatory process?*

4. *With broader representation (such as food processors and consumer groups), NABC should conduct further exploration of the relationship between the government's regulatory role, particularly the safety statutes and issues of choice such as labeling provisions.*

Albrecht, Don E.
Dept. of Rural Sociology
Texas A&M University
College Station, TX 77843 -2125

Alleger, Deanna
Dept. of Animal Science
Texas A&M University
College Station, TX 77843-2471

Arntzen, Charles J.
Texas A&M University
College Station, TX 77843

Bagnall, Brian
Smithkline Beecham
1600 Paoli Pike
West Chester, PA 19380

Bauer, Nathan
5736 Chelsea Circle
Bryan, TX 77802

Baumgardt, Bill R.
Purdue University
116 AGAD Bldg.
West Lafayette, IN 47907-1140

Bazer, Fuller W.
Dept. of Animal Science
Texas A&M University
College Station, TX 77843-2471

Bennett, Richard
University of California
Cooperative Extension
2604 Ventura Ave.
Santa Rosa, CA 95403

Benton, Charles
Texas Farm Bureau
PO Box 2689
Waco, TX 76702

Berkowitz, David
FDA
5600 Fishers Lane, Rm. 1134
Rockville, MD 20857

Black, Donald L.
University of Massachusetts
Paige Laboratory
Amherst, MA 01003

Brigl, Terrence
Environmental Growth Chambers
3700 Manchaca, #209
Austin, TX 78704

Brom, Frans W.A.
Ctr. for Bio-ethics and Health Law
Heidelberglaan2/ 3584 CS Utrecht
The Netherlands

Brown, Tony
FSIS Project Management
Texas A&M University
College Station, TX 77843-4460

Bullard, Linda
The European Parliament
rue Belliard
B-1047 Brussels, Belgium

Bullock, Bruce
University of Missouri
2-64 Agriculture Bldg.
Columbia, MO 65211

Burkhardt, Jeffrey
University of Florida
Gainesville, FL 32605

Busch, Lawrence
Dept. of Sociology
Michigan State University
East Lansing, MI 48824

Butler, Bees
Dept. of Ag Economics
University of California
Davis, CA 95616

Byers, Floyd M.
Dept. of Animal Science
Texas A&M University
College Station, TX 77843-2471

Carstens, Gordon
Dept. of Animal Science
Texas A&M University
College Station, TX 77843-2471

Cartwright, Aubrey L.
Dept. of Poultry Science
Texas A&M University
College Station, TX 77843-2472

Clancy, Kate
Syracuse University
34 Slocum Hall
Syracuse, NY 13244

Comstock, Gary
Dept. of Philosophy
Iowa State University
Ames, IA 50011

Cox, Martha / APANT
3600 Springbrook Dr.
Dallas, TX 75205

de la Concha, Andres
Texas Ag Exp. Station
7887 N. Highway 87
San Angelo, TX 76901

Cross, Russell
331 E Administration Bldg.
Washington, DC 20050

Crouch, Martha L.
Biology, Indiana University
Bloomington, IN 47405

Davenport, Manuel
Dept.of Philosophy
Texas A&M University
College Station, TX 77843-4237

Dean, Wesley
Texas A&M University
College Station, TX 77843-4355

Donis, Ruben
Dept. of Veterinary Science
University of Nebraska
Lincoln, NE 68583-0905

Dunn, Peter, E.
Entomology, Purdue University
West Lafayette, IN 47907-1158

Edwards, John
Texas A&M University
College Station, TX 77843

Ellett, E.W.
College of Veterinary Medicine
Texas A&M University
College Station, TX 77843

Erpelding, Dennis L.
Elanco Animal Health
1901 L Street NW, #705
Washington, DC 20036

Fair, Frank
Dept. of Psychology
Sam Houston State University
Huntsville, TX 77341

Fallert, Richard
USDA/ ERS / CED
1301 New York Ave., NW
Washington, DC 20005-4788

Fernandes, Antonio
Oklahoma State University
104-6 N. University Place
Stillwater, OK 74074

First, Neal L.
University of Wisconsin
752 Animal Science Bldg.
Madison, WI 53706

Fox, Michael W.
Humane Society of the U.S.
2100 L Street NW
Washington, DC 20037

Frahm, Richard R.
3007 Lancaster Drive
Blacksburg, VA 24060

Friend, Ted
Dept. of Animal Science
Texas A&M University
College Station, TX 77843-2471

Frydenlund, John
USDA Marketing and Inspection Serv.
228 W Administration Bldg.
Washington, DC 20250

Gast, Robert
Michigan State University
Ag Exp Station
109 Agriculture Hall
East Lansing, MI 48824-1039

Gastel, Barbara
Texas A&M University
College Station, TX 77843-4111

Gipson, Virginia
US FDA
3032 Bryan Street
Dallas, TX

Goodwin, Jeff
Texas A&M University
College Station, TX 77843-2124

de Grassi, Ria
California Farm Buruea
1601 Exposition Blvd.
Sacramento, CA 95815

Gray, Ian
Michigan State University
Ag Exp Station
109 Agriculture Hall
East Lansing, MI 48824-1039

Guttman, Helene N.
USDA Ag Research
BARC W, Bldg. 002, Rm 105
Beltsville, MD 20705

Hegevourt, Robert
Dept. of Animal Science
Texas A&M University
College Station, TX 77843-2471

Hand, Michael
Dept. of Philosophy
Texas A&M University
College Station, TX 77843

Hardy, Ralph W. F.
Boyce Thompson Institute
Tower Road
Ithaca, NY 14853-1801

Harlander, Susan
Land O'Lakes, Inc.
PO Box 116
Minneapolis, MN 55440

Harmon, Bud G.
Purdue University
Lilly 3-115
West Lafayette, IN 47907-1151

Harms, Paul
Dept. of Animal Science
Texas A&M University
College Station, TX 77843-2471

Hazard, Holly E.
Doris Day Animal League
900 Second Street
Washington, DC 20002

Hettinger, Edwin
College of Charleston
Charleston, SC 29424

Hill, Steve
Communications Dept.
Texas A&M University
College Station, TX 77843

Hummler, Lisa J.
Dept. of Philosophy
Texas A&M University
College Station, TX 77843-4237

Hunter, Dianna
10745 Hwy 8
Floodwood, MN 55736

Hurt, David
Nat'l Livestock & Meat Board
444 N. Michigan Avenue
Chicago, IL 60611

Ingle, John
Boyd GSRC
University of Georgia
Athens, GA 30619

Jamieson, Dale
Dept. of Philosophy
University of Colorado
Boulder, CO 80309

Jayawant, Mandar
Texas A&M University
College Station, TX 77843-4355

Johnson, Bryan H.
Maryland Ag Exp Station
1201 Symons Hall
College Park, MD 20742

Jones, Daniel D.
USDA
Office of Biotechnology
14th & Independence Ave. SW
Washington, DC 20250

Juliff, W.F.
FSIS Project Management
Texas A&M University
College Station, TX 77843-4460

Kasari, Ellen
Dept. Lab Animal Care
Texas A&M University
College Station, TX 77843-4473

Kasari, Tom
Large Animal Medicine
Texas A&M University
College Station, TX 77843-4475

King, David A.
Ag Communications, Purdue University
West Lafayette, IN 47907-1143

King, Lauriston
Univ. Research, Texas A&M University
College Station, TX 77843-3121

Klotz, Cassandra
USDA / ERS
1301 New York Ave., NW
Washington, DC 20005-4788

Kneen, Brewster
125 Highfield Road
Toronto, Ontario
Canada M4L 2V4

Kopchick, John J.
Edison Animal Biotech Center
Ohio University
Athens, OH 45701

Kimpel, Janice
Technology Transfer
Boyd GSRC
University of Georgia
Athens, GA 30602

Kriewaldt, David
Texas A&M University
College Station, TX 77843-4355

Kunkel, H.O.
Animal Sci., Texas A&M University
College Station, TX 77843-2471

Kutach, Douglas
Texas A&M University
College Station, TX 77843-4355

Leach, Laurie
Animal Health Institute
119 Oronoco Steet
Alexandria, VA 22314-2058

Lee, J. Charles
Texas Ag Exp Station
Texas A&M University
College Station, TX 77843-2142

Lee, Patricia P.
International Livestock Congress
Texas A&M University
College Station, TX 77843-2471

Love, Todd
Animal Nutrition
Texas A&M University
College Station, TX 77843-2471

Linskins, Michael
Dutch Society for the Protection
of Animals
PO Box 85g80
The Netherlands

MacDonald, June Fessenden
NABC / 159 Biotechnology Bldg.
Cornell University
Ithaca, NY 14853-2703

MacKenzie, David R.
USDA / CSRS
901 D Street, SW
Washington, DC 20250-2200

McGregor, Martin L.
Baker & Botts
910 Louisiana
Houston, TX 77002-4995

Magill, Jane
Dept. of Biochemistry
Texas A&M University
College Station, TX 77843-2132

Malloy, Donna L.
USDA / APHIS
6505 Belcrest Road
Hyattsville, MD 20782

Martin, Marshall A.
Ctr for Ag Policy &Tech.Assessment
Purdue University
West Lafayette, IN 47907-1151

Meagher, Laura
Ag Biotechnology Center
Rutgers University
New Brunswick, NJ 08903

Meeker, David
Nat'l Pork Producers Council
Box 10383
Des Moines, IA 50306

Mellon, Margaret
Nat'l Biotechnology Policy Ctr.
Nat'l Wildlife Federation
1400 16th St., NW
Washington, DC 20036

Merrifield, Robert G.
Ag Exp St., Texas A&M University
College Station, TX 77843-2147

Murphy, George
Biotransplants
Bldg. 96, 13th Street
Charleston Navy Yard, MA 02129

Murphy, Joan
Food Chemical News
1101 Pennsylvania, SE
Washington, DC 20003

Murray, Jim
University of California
Davis, CA 95616

Nelkin, Dorothy
Dept. of Sociology
New York University
New York, NY 10003

Nelson, Darrell W.
University of Nebraska
207 Agriculture Hall
Lincoln, NE 68583-0704

Nicholas, Robert B.
McDermott, Will & Emery
1850 K Street, NW Suite 500
Washington, DC 20006

Offutt, Susan E.
Nat'l Research Council
2191 Constitution Ave., NW
Washington, DC 20418

O'Hara, Kate
NABC / 159 Biotechnology Bldg.
Cornell University
Ithaca, NY 14853-2703

Oltjen, Robert R.
Animal Sciences Nat'l Program
005 BARC W, R. 134
Beltsville, MD 20705

Osburn, Bennie
Veterinary Medicine
University of California
Davis, CA 95616

Ott, Troy L.
Dept. of Animal Science
Texas A&M University
College Station, TX 77843-2471

Patel, Bharat L.
USDA / FSIS; Science and Technology
Rm. 4911 South Building
Washington, DC 20250

Piedrahita, Jorge, A.
Dept. of Veterinary Medicine
Texas A&M University
College Station, TX 77843

Plummer, Bill
Animal Sciences & Industry
Cal Poly State University
San Luis Obispo, CA 93407

Pomp, Daniel
Dept. of Animal Science
Oklahoma State University
Stillwater, OK 74078

Poos, Mary
Nat'l Research Council
2101 Constitution Ave., NW
Washington, DC 20418

Ritter, Ellen
Ag Communications
229 Reed McDonald
Texas A&M University
College Station, TX 77843

Rollin, Bernard E.
Dept. of Philosophy
Colorado State University
Fort Collins, CO 80523

Rothschild, Max F.
Iowa State University
18 Curtiss Hall
Ames, IA 50011

Rowan, Andrew
Center for Animals and Public Policy
Tufts University
Westboro, MA 01536

Seetharaman, Koushik
Dept. of Soil and Crop Science
Texas A&M University
College Station, TX 77843-2477

Seidel, George E. Jr.
Animal Repoduction and
Biotechnology
Colorado State University
Fort Collins, CO 80523

Shadduck, John
Texas A&M University
College Station, TX 77843

Shaw, Eric T.
Texas A&M University
College Station, TX 77843-4355

Shore, Scott
NC Dept. of Agriculture
PO Box 27647
Raleigh, NC 27611

Siddiq, Asif
Texas A&M University
College Station, TX 77843-4355

Sigurdson, Chris
Ag Communications
Purdue University
West Lafayette, IN 47907

Stabinsky, Doreen
Dept. of Plant Pathology
University of California
Davis, CA 95616

Steele, Scott
Texas A&M University
College Station, TX 77843-4355

Stenholm, Rep. Charles W.
1301 Longworth Office Bldg.
Washington, DC 20515

Sullivan, John J.
American Breeders Service
6908 River Road
DeForest, WI 53532

Swan, Patricia B.
Iowa State University
107 Beardshear Hall
Ames, IA 50011

Terry, Martin
Scientific Activities, AHI
119 Oronoco Street
Alexandria, VA 22314-2058

Thompson, Paul B.
Ctr. for Biotechnology Policyand Ethics
Texas A&M University
College Station, TX 77843

Truitt, Jay H.
Missouri's Farm Voice
102 N. Mason
Carrollton, MO 64633

Turk, Danny
Dept. of Animal Science
Texas A&M University
College Station, TX 77843-2471

Turner, Nancy
Dept. of Animal Science
Texas A&M University
College Station, TX 77843-2471

Urban, Oto
USDA / FSIS
14th & Independence Ave,
Washington, DC 20250

Vaid, Joyotsna
Dept. of Psychology
Texas A&M University
College Station, TX 77843-4235

Varner, Gary
Ctr. for Biotechnology Policy
and Ethics
Texas A&M University
College Station, TX 77843-4355

van Vugt, Frits
Ministry of Agriculture
PostBus 20401
2500 EK the Hague
The Netherlands

Voichick, Jane
Dept. of Nutritional Science
University of Wisconsin
1415 Linden Drive
Madison, WI 53705

Walker, William H. III
Agricenter International
7777 Walnut Grove Raod
Memphis, TN 38120

Westhusin, Mark
Dept. of Physiology
Texas A&M University
College Station, TX 77843-4466

Wild, Jim
Dept. of Biochemistry
Texas A&M University
College Station, TX 77843-2128

Winhager, Steven
Texas A&M University
College Station, TX 77843-4355

Youngs, Curt
Dept. of Animal Science
Iowa State University
Ames, IA 50011

Zalesky, Doug
LSU Agricultural Center
226 Knapp Hall
Baton Rouge, LA 70803-1900

Zimbelman, Robert G.
American Society of Animal Science
Bethesda, MD 20814

Zinnen, Thomas M.
University of Wisconsin
Biotechnology Center
1710 University Avenue
Madison, WI 53705